GREATMAPS

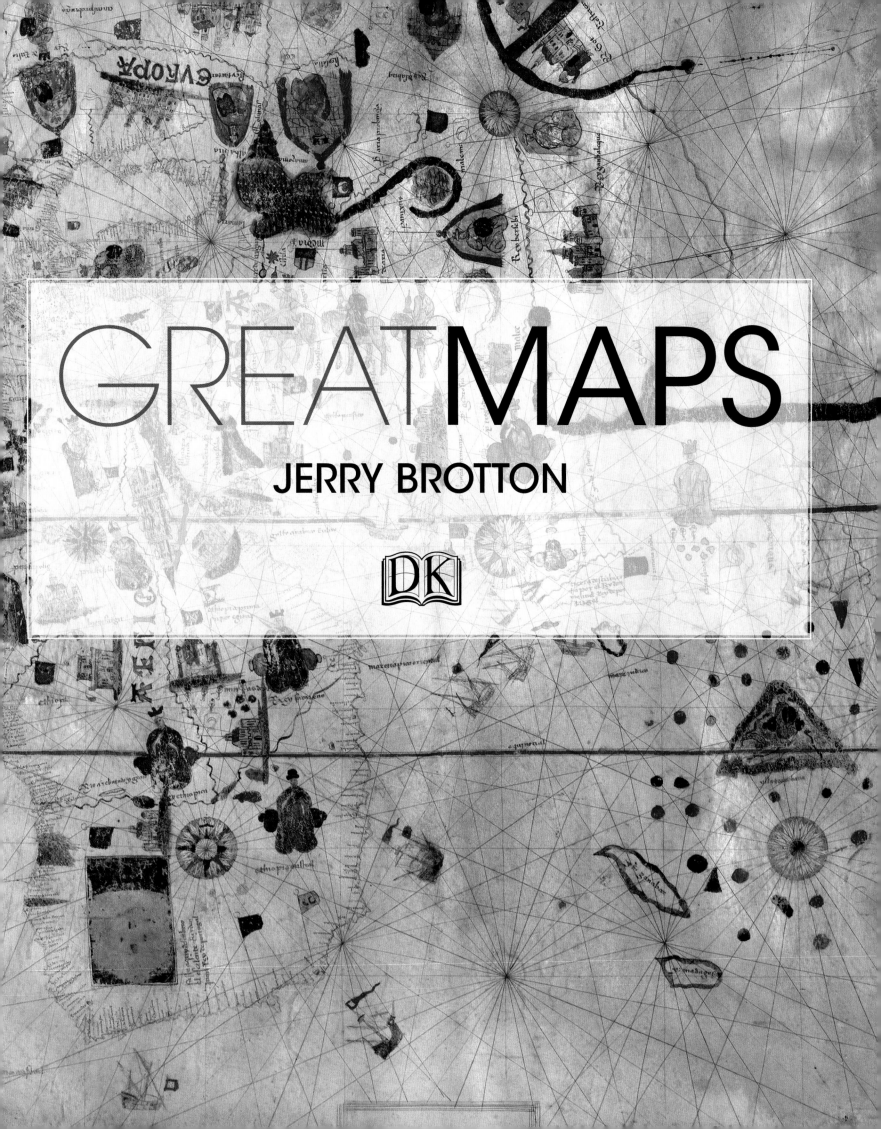

GREATMAPS

JERRY BROTTON

DK

LONDON, NEW YORK, MUNICH,
MELBOURNE, DELHI

Editorial Team	Catherine Saunders, Hugo Wilkinson
Editorial Assistant	Stuart Neilson
Senior Art Editor	Gillian Andrews
Senior Designer	Stephen Bere
Producer, Pre-Production	Lucy Sims
Senior Producer	Mandy Inness
Picture Research	Roland Smithies, Sarah Smithies
Jacket Designer	Laura Brim
Jacket Editor	Maud Whatley
Managing Editor	Stephanie Farrow
Senior Managing Art Editor	Lee Griffiths
Publisher	Andrew Macintyre
Art Director	Phil Ormerod
Publishing Director	Jonathan Metcalf

DK INDIA

Editor	Esha Banerjee
Art Editor	Pooja Pipil
Assistant Art Editor	Tanvi Sahu
DTP Designers	Shankar Prasad, Vijay Kandwal
Managing Editor	Kingshuk Ghoshal
Managing Art Editor	Govind Mittal
Pre-production Manager	Balwant Singh
Production Manager	Pankaj Sharma

First published in Great Britain
in 2014 by
Dorling Kindersley Limited
80 Strand, London WC2R 0RL
A Penguin Random House Company

2 4 6 8 10 9 7 5 3 1
001-193222-September/2014

Copyright © 2014 Dorling Kindersley Limited

All rights reserved. No part of this publication may
be reproduced, stored in a retrieval system, or
transmitted in any form or by any means, electronic,
mechanical, photocopying, recording, or otherwise
without the prior written permission of the
copyright owners.

A CIP catalogue record for this book is
available from the British Library
ISBN 978-1-4093-4571-8

Printed and bound in Hong Kong

Discover more at **www.dk.com**

AUTHOR'S NOTE
**This book is dedicated to my father,
Alan Brotton**

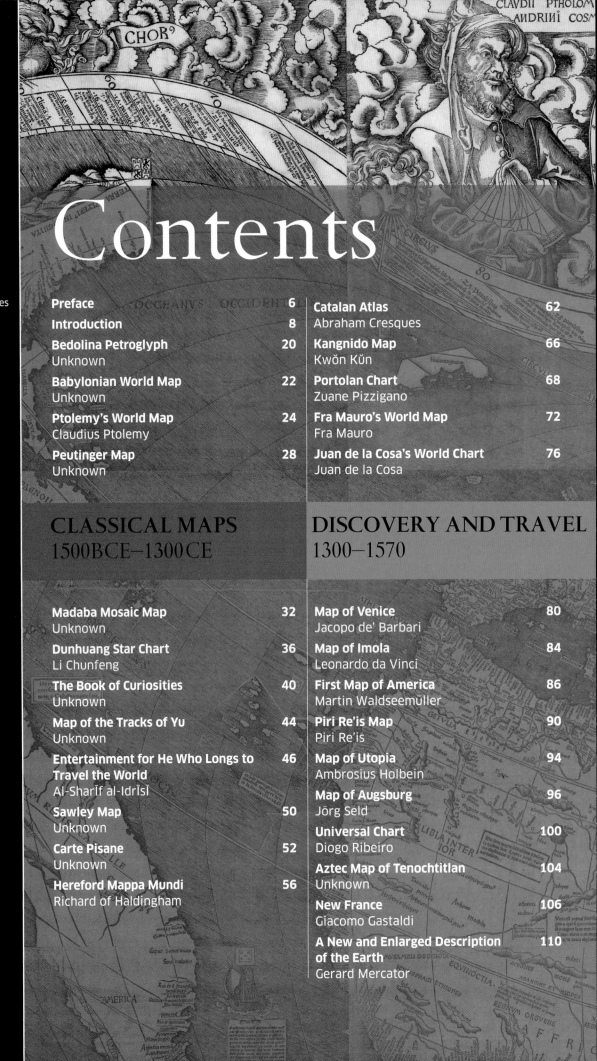

Contents

Preface	6
Introduction	8
Bedolina Petroglyph Unknown	20
Babylonian World Map Unknown	22
Ptolemy's World Map Claudius Ptolemy	24
Peutinger Map Unknown	28

Catalan Atlas Abraham Cresques	62
Kangnido Map Kwŏn Kŭn	66
Portolan Chart Zuane Pizzigano	68
Fra Mauro's World Map Fra Mauro	72
Juan de la Cosa's World Chart Juan de la Cosa	76

CLASSICAL MAPS
1500 BCE–1300 CE

DISCOVERY AND TRAVEL
1300–1570

Madaba Mosaic Map Unknown	32
Dunhuang Star Chart Li Chunfeng	36
The Book of Curiosities Unknown	40
Map of the Tracks of Yu Unknown	44
Entertainment for He Who Longs to Travel the World Al-Sharīf al-Idrīsī	46
Sawley Map Unknown	50
Carte Pisane Unknown	52
Hereford Mappa Mundi Richard of Haldingham	56

Map of Venice Jacopo de' Barbari	80
Map of Imola Leonardo da Vinci	84
First Map of America Martin Waldseemüller	86
Piri Re'is Map Piri Re'is	90
Map of Utopia Ambrosius Holbein	94
Map of Augsburg Jörg Seld	96
Universal Chart Diogo Ribeiro	100
Aztec Map of Tenochtitlan Unknown	104
New France Giacomo Gastaldi	106
A New and Enlarged Description of the Earth Gerard Mercator	110

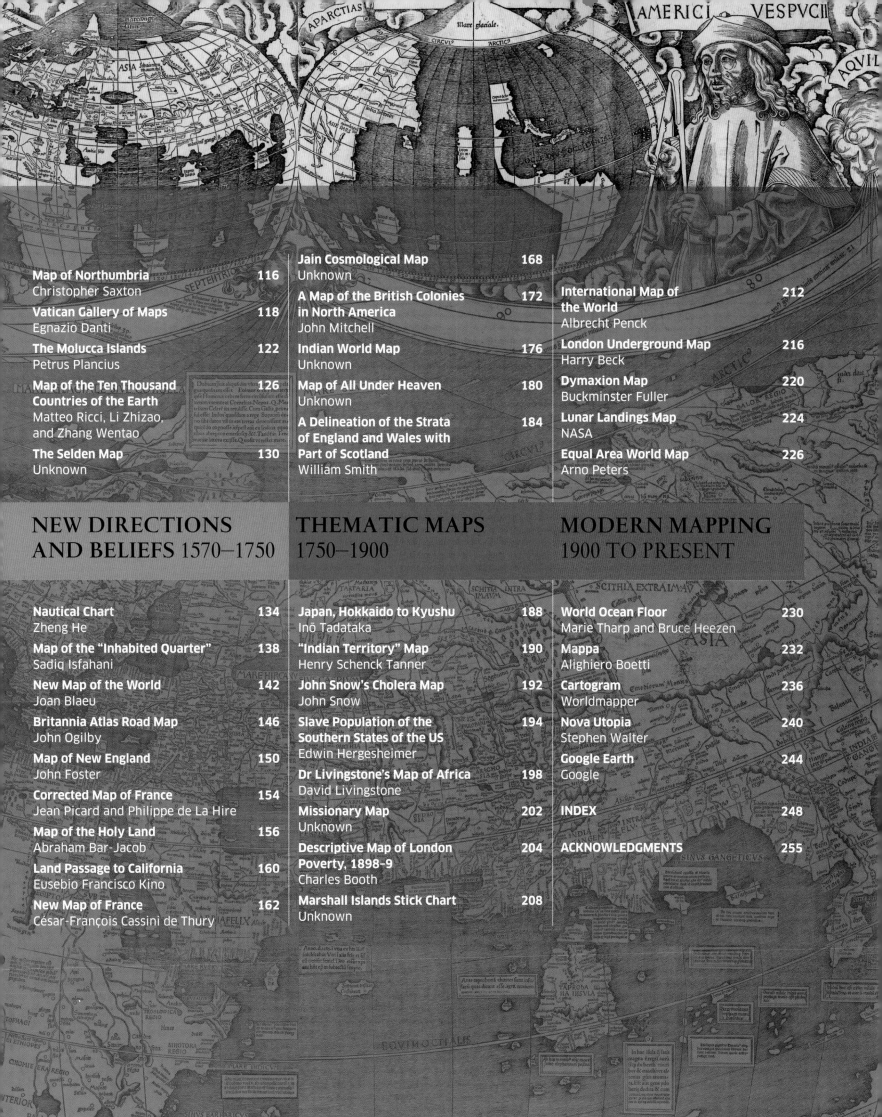

Map of Northumbria 116
Christopher Saxton

Vatican Gallery of Maps 118
Egnazio Danti

The Molucca Islands 122
Petrus Plancius

Map of the Ten Thousand 126
Countries of the Earth
Matteo Ricci, Li Zhizao,
and Zhang Wentao

The Selden Map 130
Unknown

Jain Cosmological Map 168
Unknown

A Map of the British Colonies 172
in North America
John Mitchell

Indian World Map 176
Unknown

Map of All Under Heaven 180
Unknown

A Delineation of the Strata 184
of England and Wales with
Part of Scotland
William Smith

International Map of 212
the World
Albrecht Penck

London Underground Map 216
Harry Beck

Dymaxion Map 220
Buckminster Fuller

Lunar Landings Map 224
NASA

Equal Area World Map 226
Arno Peters

NEW DIRECTIONS AND BELIEFS 1570–1750

THEMATIC MAPS 1750–1900

MODERN MAPPING 1900 TO PRESENT

Nautical Chart 134
Zheng He

Map of the "Inhabited Quarter" 138
Sadiq Isfahani

New Map of the World 142
Joan Blaeu

Britannia Atlas Road Map 146
John Ogilby

Map of New England 150
John Foster

Corrected Map of France 154
Jean Picard and Philippe de La Hire

Map of the Holy Land 156
Abraham Bar-Jacob

Land Passage to California 160
Eusebio Francisco Kino

New Map of France 162
César-François Cassini de Thury

Japan, Hokkaido to Kyushu 188
Inō Tadataka

"Indian Territory" Map 190
Henry Schenck Tanner

John Snow's Cholera Map 192
John Snow

Slave Population of the 194
Southern States of the US
Edwin Hergesheimer

Dr Livingstone's Map of Africa 198
David Livingstone

Missionary Map 202
Unknown

Descriptive Map of London 204
Poverty, 1898–9
Charles Booth

Marshall Islands Stick Chart 208
Unknown

World Ocean Floor 230
Marie Tharp and Bruce Heezen

Mappa 232
Alighiero Boetti

Cartogram 236
Worldmapper

Nova Utopia 240
Stephen Walter

Google Earth 244
Google

INDEX 248

ACKNOWLEDGMENTS 255

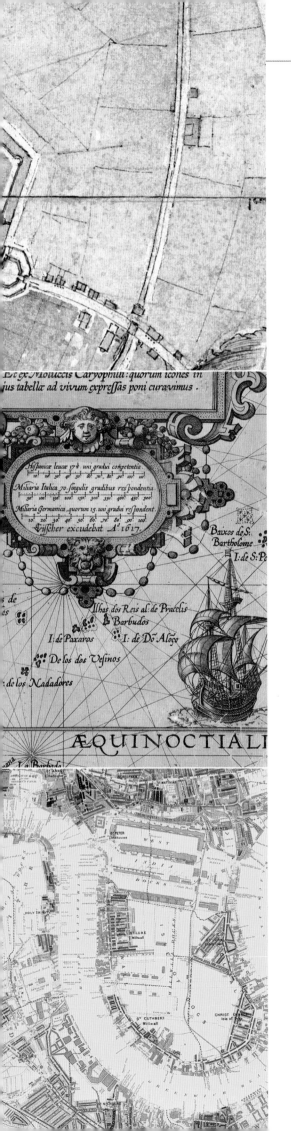

Preface

Today, maps are regarded primarily as locational or navigational tools. Made of paper or, more likely, accessed digitally, they provide information about our surroundings or guide us from one place to another with maximum speed and efficiency. However, throughout history, maps have served a variety of purposes. In fact, ever since mankind first learned how to make graphic marks on rock up to 40,000 years ago, people have created maps as a way of conceptualizing themselves in relation to their environment. Thus, maps are as much about existence as they are about orientation. Processing our surroundings spatially is a basic human activity that psychologists call "cognitive mapping". While other animals demarcate their territories, we are the only species capable of mapping ours.

MAKING SENSE OF THE WORLD

So, what is a map? The word "map" was first used in English in the 16th century and, despite more than 300 subsequent, competing interpretations of the term, most scholars now broadly agree that a map can be defined as "a graphic representation that presents a spatial understanding of things, concepts, or events in the human world". Although this definition might seem vague, it frees maps from the constraints of being considered merely scientific tools, and allows an extraordinary array of renditions to co-exist under the title of "maps" – celestial, astrological, topographical, theological, spiritual, statistical, political, navigational, imaginative, and artistic. Such a broad definition also embraces the sheer variety of different cultural traditions, including what, how, and why different communities have made and do make maps, from the Greek *pinax* and the Latin *mappa*, to the Chinese *tu* and the Arabic *ṣūrah*. Many of these different traditions are described in the following pages, which start with a map carved in stone more than 3,500 years ago, and then explore examples made using clay, mosaic, papyrus, animal skin, paper, and electronic media.

  Great Maps features a diverse selection of maps made during key moments in world history, and explains how they provide important answers to the most urgent questions of their eras. This book presents mapmaking as a truly global phenomenon – it is an activity common to every race, culture, and creed, although each one has very a distinctive way of mapping its particular world. The book also reveals that, despite the claims of many mapmakers throughout history, there is no such thing as a perfect map. Maps are always subjective, and there are invariably many different ways of mapping the same area. *Great Maps* is broken down into five chronological chapters, but this does not mean that the history of maps is one that becomes gradually more scientifically accurate and "correct" as we approach the modern age. Instead, each section explains how maps answer the specific needs of their intended audience, so that a 13th-century religious map that puts Jerusalem at its centre is as "true" to its original audience as the digital maps many of us regularly consult on our mobile phones today.

FAVONIVS.

P

RICVS

A

R. AFFRICVS.

IVDECA

O

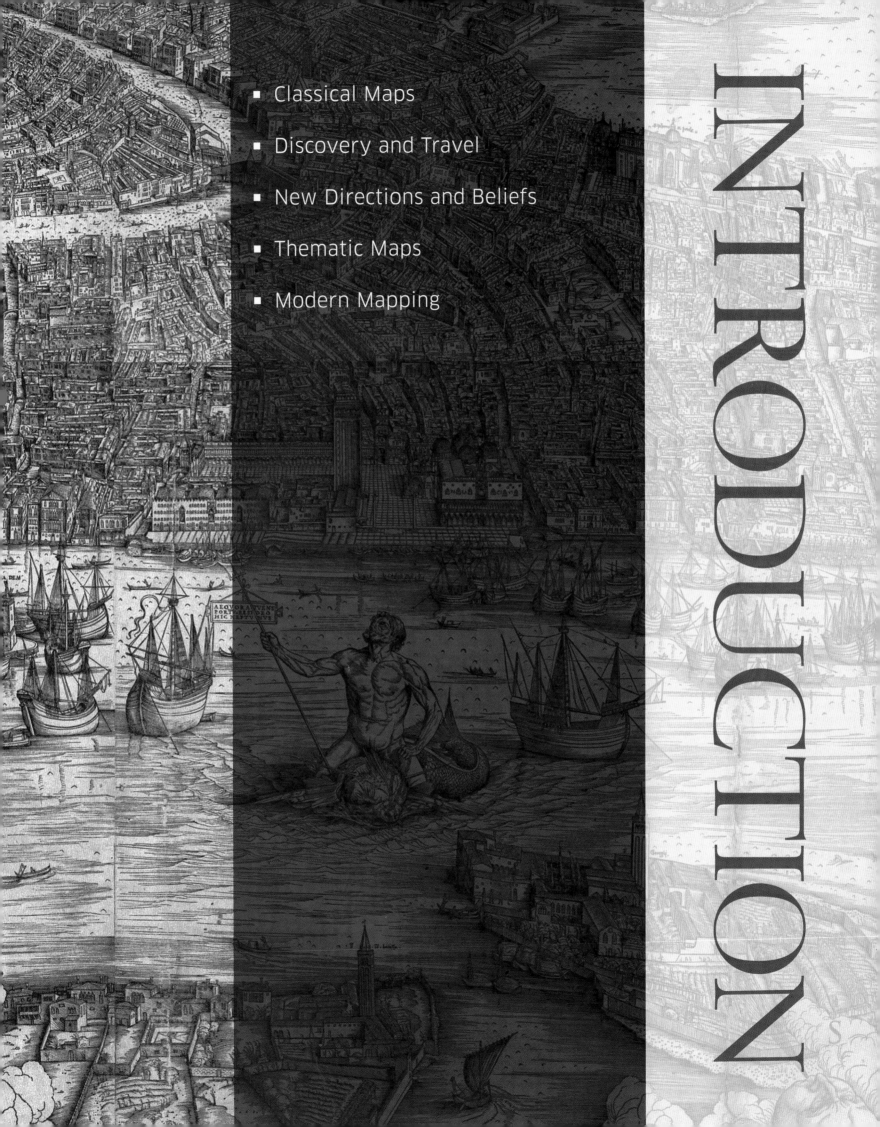

- Classical Maps
- Discovery and Travel
- New Directions and Beliefs
- Thematic Maps
- Modern Mapping

INTRODUCTION

Classical Maps

In the classical world, which comprised the pre-Christian Babylonian, Persian, Greek, and Roman empires, maps fulfilled a variety of functions – celebrating imperial world views, explaining creation, describing the heavens, or visualizing religious beliefs. Their use in travel itineraries and navigation developed much later and was often secondary, because their scale and detail were initially limited. One of the difficulties scholars face when assessing maps from the classical era is that there are very few still in existence. Surviving written accounts reveal that world and regional maps were being drawn as early as the 7th century BCE. Books known as *Periodos Gēs* ("Circuit of the Earth") included maps showing the world as a flat disc surrounded by water, encompassing Europe, Asia, and Africa.

By the Platonic era (427–347 BCE), the Greeks knew that the Earth was spherical, and Aristotle (384–322 BCE) argued that it could also be divided into different *klimata* (climes). By the third century BCE, Alexandria in Egypt had become a renowned centre for geographical study, thanks to its fabled library. Eratosthenes (c.275–c.194 BCE), the chief librarian, wrote one of the first of several books to be entitled *Geography,* in which he described and mapped the entire known world. Eratosthenes used

mathematics and geometry to draw his maps of the Earth. He also calculated the planet's circumference for the first time and, amazingly, was within 2,000–4,000km (1,250–2,500 miles) of its correct measurement. Later, Eratosthenes's scientific methods were adopted by Claudius Ptolemy, whose own book *Geography* (150 CE) listed the coordinates of 8,000 places in the Greco-Roman world, and explained how to use geometry and mathematics to insert them into a grid of latitude and longitude known as a graticule.

BEYOND SCIENCE

Elsewhere, other mapmakers had little interest in following the Greek tradition and using science to make their point. The *Peutinger Map*, for example, is a long, distorted survey of the Roman Empire as it began to fragment in the 4th century CE. Religious maps such as the Madaba mosaic and the medieval mappae mundi use theology rather than geometry to orient themselves. In China, astrological charts such as the *Dunhuang Star Chart* try to show how the movement of the heavens above affected mankind below. Others, such as the *Map of the Tracks of Yu* do offer a measured grid within which to map the empire, but one that used very different

▲ **PTOLEMY'S WORLD MAP** Ptolemy plotted the *ecumene* (known world), which for him stretched from the Canary Islands in the west to Korea in the east, with a huge Mediterranean Sea and Indian Ocean in the middle.

When the **observer looks at these maps** and these countries explained, he sees **a true description** and pleasing form

AL-SHARIF AL-IDRISI, *ENTERTAINMENT FOR HE WHO LONGS TO TRAVEL THE WORLD*

calculations to those developed by the Greeks. Many classical maps were created solely to be used by the elite – emperors, scholars, and holy men – but others were made for more practical purposes. Across the Mediterranean Sea, both Muslims and Christians were making portolan charts, navigational maps with no agenda other than ensuring a safe voyage. Usually unadorned and concentrating on drawing and naming coastlines, these maps were used by merchants and pilots whose livelihoods and future prosperity depended on them getting from one place to another quickly and safely. These types of maps, such as those in the Egyptian *Book of Curiosities* and the earliest surviving example of a portolan chart, the *Carte Pisane*, pointed cartography in a new direction.

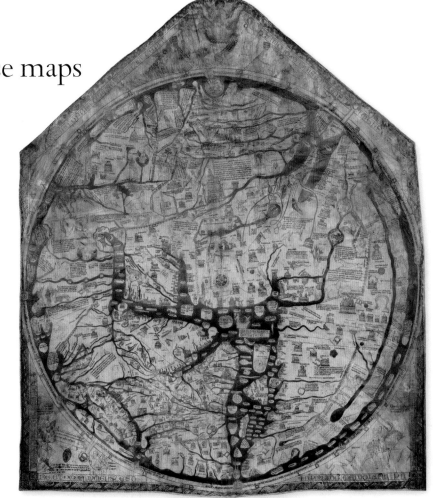

▲ **HEREFORD MAPPA MUNDI** The original function of this map is unknown, but the inscriptions relate to medieval geography, theology, cosmology, and zoology.

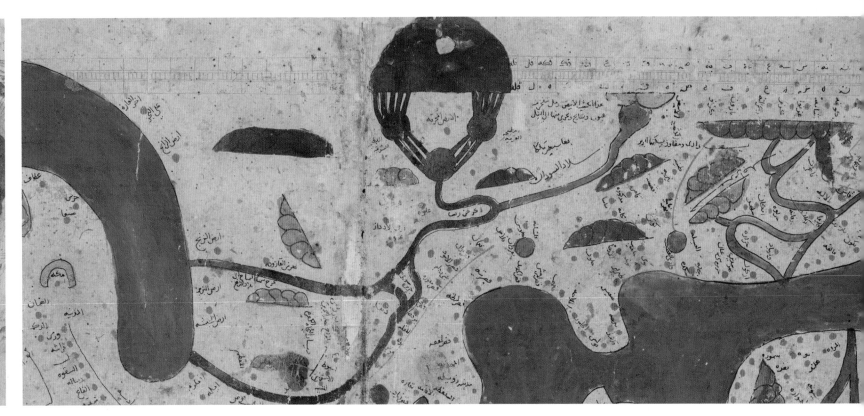

▲ **THE BOOK OF CURIOSITIES** This map combines practical navigation uses with an Islamic world view. South is at the top and both the Arabian Peninsula and Mecca are prominently featured.

Discovery and Travel

The beautifully illustrated but fantastical *Catalan Atlas* depicts an inhabited world in 1375 that stretches from Portugal to China and is full of bizarre creatures and mythical lands. However, less than 200 years later, Gerard Mercator's *New and Enlarged Description of the Earth* represents the world as we understand it today. Remarkable changes took place between the late 14th century and the late 16th century – so much so that the period is often known as the "Age of Discovery" – and yet many of them might never have happened without some of the maps discussed in this book. As European, Ottoman, and Chinese explorers began to travel to places beyond their realms, they required maps – firstly to help them get there (usually by sea) and then to show off their overseas possessions once they had claimed them.

To make the Portuguese maritime expeditions out into the northern Atlantic possible in the early 15th century, explorers had to modify the Mediterranean portolan sailing charts and navigate using the stars. The resulting charts include Zuane Pizzigano's portolan chart of 1424. Apparently straightforward, on closer inspection it

appears to show a collection of "phantom" islands, which historians speculate may represent the Caribbean, or even the Americas – 70 years before their official "discovery" by Christopher Columbus. Over in Korea, Kwŏn Kŭn also offered some surprisingly accurate details. His *Kangnido* map seems to depict a circumnavigable southern Africa at least 80 years before the Portuguese took credit for discovering it.

MARITIME AND PHILOSOPHICAL MAPS

This was a period in which, despite great innovations in the science of mapmaking, long-distance maritime travel remained hazardous and maps were often disorienting or distorted. Juan de la Cosa produced the first map showing Columbus's landfall in the Americas, but he uses two completely different scales for the Old and New Worlds. Diogo Ribeiro's universal chart, on the other hand, is one of the earliest examples of cartographic propaganda, manipulating the position of the Moluccas spice islands to promote Spain's claims to them. Only the western half of a

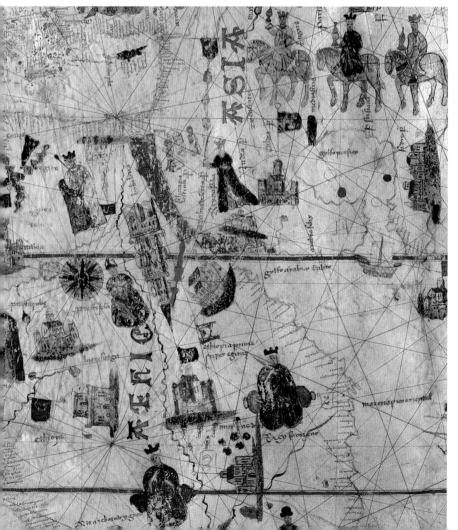

◀ **JUAN DE LA COSA'S WORLD CHART** The eastern part of the map uses Ptolemy's scale to represent Europe, Africa, and Asia, while the western half uses a much larger scale to show the Americas.

mysterious world map created by Ottoman admiral Piri Re'is remains. It presented a projection that still baffles modern cartographers.

The discovery of these new worlds also inspired more philosophical maps. The publication of Thomas More's book *Utopia* in 1516 was partly the result of Spanish exploration in the Americas, and it generated a new genre of utopian writing that described an island populated by an ideal society. Ambrosius Holbein's map, created to illustrate More's book, is the first in a long line of maps of this famous, fictional island, continuing all the way to British artist Stephen Walter and his 2013 work, *Nova Utopia*.

EMBRACING SCIENCE

What unites many of the practical maps of this era is the desire to use new scientific methods to measure the space they represent. In Europe, Ptolemy's legacy led mapmakers to turn gradually away from theology, and instead to use geometric map projections and regional surveying methods to create maps that

> ## Maps codify the miracle of existence
>
> **NICHOLAS CRANE**, *MERCATOR: THE MAN WHO MAPPED THE PLANET*

helped their owners to conquer, rule, or defend the territories they depicted. This desire can also be identified in early 16th-century city maps of Venice and Augsburg, Leonardo da Vinci's beautiful map of Imola (even though it was a map of military strategy designed to repel potential invaders), and the Aztec map of Tenochtitlan, which captures an urban geometry particular to the pre-Columbian Mexica people just before the destruction of the Aztec Empire. If this was a period when maps showed new worlds and ideas, it was also a time when the art of the mapmaker was beginning to be seen as a useful tool in wielding political, military, and economic power, both at home and abroad.

▲ **FIRST MAP OF AMERICA** Despite spanning 12 sheets, Martin Waldseemüller's map struggles to contain all the latest discoveries, giving it a strange, bulbous shape that threatens to break out of its frame.

▲ **AZTEC MAP OF TENOCHTITLAN** Maps created by the Aztecs, such as this one, have a unique cartographic style, using a mixture of geometry, hieroglyphics, and mythical imagery to present their world view.

New Directions and Beliefs

As the world map that we now recognize came slowly into focus during the 17th century, cartography became increasingly specialized. Thanks to developments in printing, maps could also reach a wider audience. As a result, they became a more commercial enterprise, employing teams of craftsmen and scholars, who were commissioned by various interest groups to represent their beliefs. In Britain, John Ogilby saw the map's potential as a road atlas, while across the English Channel, the Cassini family drew on French royal patronage to develop new scientific techniques in surveying. The Cassinis' *New Map of France* was not only the first comprehensive survey of a nation, predating the earliest British Ordnance Survey maps by decades, but it was also popular with French republicans – both during and after the 1789 revolution – who "nationalized" the map and promoted it as a symbol of their nation, free of monarchy.

THE CHANGING ROLE OF MAPS

Others took the map's growing power in different directions. Matteo Ricci's imposing world map is a remarkable fusion of east–west cartographic knowledge, although the map's primary function was to persuade the Chinese to adopt a Christian God. Meanwhile, the proselyte Abraham Bar-Jacob produced a Holy Land map designed for use in the Jewish Passover Seder celebrations. The diffusion of scientific mapping methods reached India and fused with south Asian mapping traditions to produce beautiful maps of the "inhabited quarter" of the world, as seen by Mughal Indian scholars, such as Sadiq Isfahani. These new cartographic techniques were also exported westwards, enabling colonists in previously uncharted places, such as North America, to survey and publish their own maps. John Foster's crude but highly effective map of puritan settlements focuses on the eastern side of America, while on the other side of the continent the Jesuit Eusebio Kino was able to prove once and for all that California was not an island. As the world became more explicable, it began to shrink, and the mapmaker's role changed from that of a savant to that of an administrator, slowly filling in the blanks across the Earth's surface.

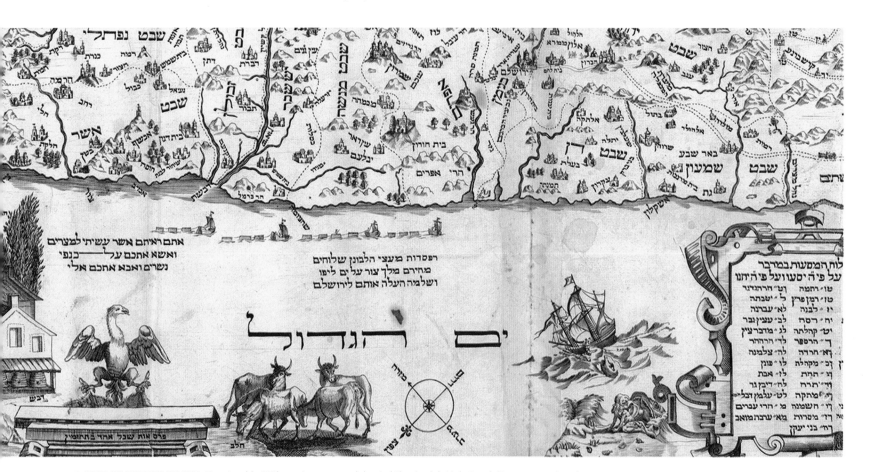

▲ **MAP OF THE HOLY LAND** Bar-Jacob's 17th-century map celebrated the Jewish Holy Land. It was reproduced in a prayer book and had little interest in enabling travellers to actually find their way across the region.

Thematic Maps

The 19th century was the great age of thematic mapping, particularly in the Western world; this was the era in which new ways of capturing and representing statistical information could be combined with the map's graphic design to create a powerful visual statement. The widespread introduction of national censuses at the beginning of the century provided mapmakers with a torrent of data on a variety of subjects, from poverty, wealth, and disease to race, religion, and slavery. New developments in design and distribution also brought maps to ever wider audiences, as techniques such as lithographic printing allowed them to be published in colour, in a fraction of the time and at much lower cost than traditional engraving methods.

MAPS AND SOCIAL CHANGE

Increasingly, maps were also being used by governments and various learned institutions to advance all kinds of political and social engineering schemes. In 1815, as William Smith proposed the first national geological map in Britain, Inō Tadataka was painstakingly mapping the entire Japanese coastline for the Tokugawa shogunate.

Just a decade later, Henry Schenck Tanner produced maps that were used by the US Congress to help facilitate the "relocation" of Native American tribes. During this period, European mapping methods were also used to disseminate the values of Western religion and culture across the globe, especially in places such as Africa. Dr David Livingstone's seemingly innocent map of Africa belies his aggressive attempt to introduce commerce and Christianity to the so-called "dark continent", while the explicitly evangelical *Missionary Map* put forth American preacher William Miller's Adventist beliefs. These, and many other contemporary maps, presumed to lay claim to people and territories over which they had no natural rights.

However, while some mapmakers simply looked to impose their value systems upon the world, others pursued more progressive agendas. Edwin Hergesheimer's map of the *Slave Population of the Southern States of the US* provided a shocking graphic that was adopted by US President Abraham Lincoln when he campaigned for the abolition of slavery during the American Civil War (1861–65). In 19th-century Britain, social policy was influenced by the thematic cartography of reformers such as John Snow, whose cholera map transformed epidemiology, and Charles Booth, who applied wealth data to a map of London, and discovered that the true extent of poverty was far greater than he had imagined. In places such as India and Korea maps celebrating themes of faith and myth continued to be made, happily coexisting alongside their more scientific counterparts.

▲ **LONDON POVERTY MAP** Social campaigner Charles Booth used a colour-coded system to present his poverty data clearly and effectively.

Neither the historian nor the cartographer can ever **reproduce the reality** they are trying to communicate

CRANE BRINTON, AMERICAN HISTORIAN

Modern Mapping

In the 20th and 21st centuries, cartography has travelled in many different directions. Although the world has fewer mysteries, the scope for mapmaking has not diminished, and some truly great maps have been created during this period. However, many of them have been criticized for presenting deeply ideological views of the modern world, or, worse, have been usurped by others in order to promote their own agendas. In fact, modern cartographers are often confronted with a troublesome paradox: they have proved time and again that any map is merely a selective and partial depiction of a territory, but this very fact makes their work a powerful tool, and therefore subject to appropriation by those eager to use maps for military, political, ideological, or propaganda purposes.

POWER AND ART

Albrecht Penck tried and failed to produce a uniform map of the whole world on a scale of 1:1,000,000. In the end, he could only watch helplessly as generals took over his project and used it for military purposes during World War I. Meanwhile, radicals and visionaries such as Arno Peters and Buckminster Fuller proposed completely new projections for mapping the world, rejecting traditional views that they regarded as Eurocentric or lacking in environmental awareness. This has led to the creation of some truly weird and wonderful maps. Indeed, Fuller's *Dymaxion Map* and Danny Dorling's series of cartograms prove that the most "accurate" maps are often those that appear the most distorted. It is hardly surprising that these innovations have attracted artists to the world of maps. Alighiero Boetti and Stephen Walter are just two of the artists who have been inspired to play with the endless possibilities of colour, shape, language, and politics of maps in their work.

20TH CENTURY PROGRESS

By the late 1960s, mankind could travel virtually anywhere, so little stood in the way of mapmakers. As a result, cartographers have been able to map a

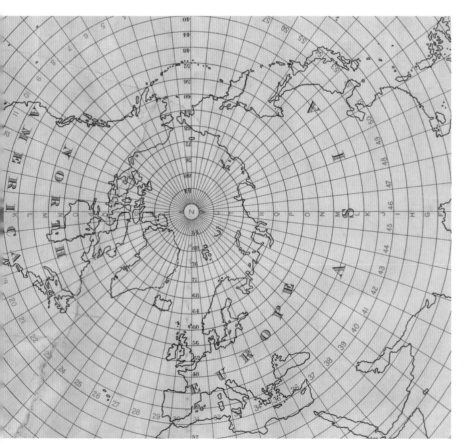

▲ **THE "MILLIONTH MAP"** Penck's vision for the *International Map of the World* involved every nation creating their own map to the same scale. It was an ambitious project and, ultimately, a doomed one.

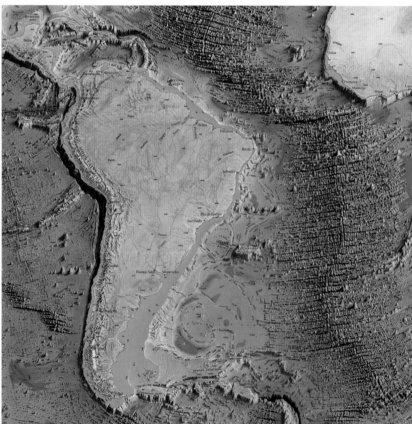

▲ **WORLD OCEAN FLOOR** The first map of the entire ocean floor was only created in 1977, thanks to US geologists Marie Tharp and Bruce Heezen, and Austrian landscape painter Heinrich Berann.

huge range of places, from the depths of the ocean to far-flung extraterrestrial worlds. In the late 1970s, geologists Marie Tharp and Bruce Heezen not only extensively mapped the ocean floor for the first time, but were also able to prove the veracity of continental drift once and for all. At the other end of the map spectrum, NASA achieved something that scientists such as Galileo and Riccioli could only have dreamed of in the 17th century, when they finally mapped the Moon.

THE DIGITAL AGE

In the 21st century, maps have swapped paper for pixels and moved into the digital age. The development of computer-generated Geographical Information Systems (GIS) since the 1960s has also given scientists, rather than cartographers, the ability to map the whole world in unprecedented detail. This has culminated in what internet service providers such as Google now call "geospatial applications". Thanks to hitherto unimaginable amounts of geographical data, any

> You can't create a perfect map. You never will

DANNY DORLING, BRITISH SOCIAL GEOGRAPHER

computer or mobile phone user can now zoom across the planet in seconds, accessing maps at extraordinary levels of resolution.

THE FUTURE OF MAPS

While digital applications signal a whole new era of mapping – one that surely marks the twilight of paper maps – many of the issues facing modern mapmakers are the same as those that confronted the Babylonians, more than 2,500 years ago. What should go on the map, and what should be left out? Who pays for a map, and who will use it? Regardless of the medium, it would seem that great maps will always be necessary. After all, they provide answers to humanity's most enduring questions: "Where am I?" and even, "Who am I?"

▲ **LUNAR LANDINGS MAP** Humans had been trying to map the Moon since the 17th century, but finally succeeded in 1969. Although mankind finally reached the Moon in that year, the map was created using images from observatories.

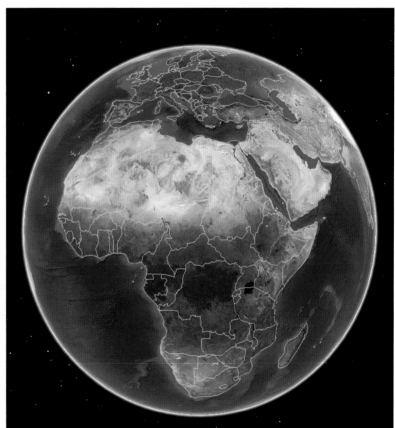

▲ **GOOGLE EARTH** Users can zoom in on their country, town, and even their street, on these high-resolution, digital geospatial maps, which can be accessed via computers, tablets, and mobile phones.

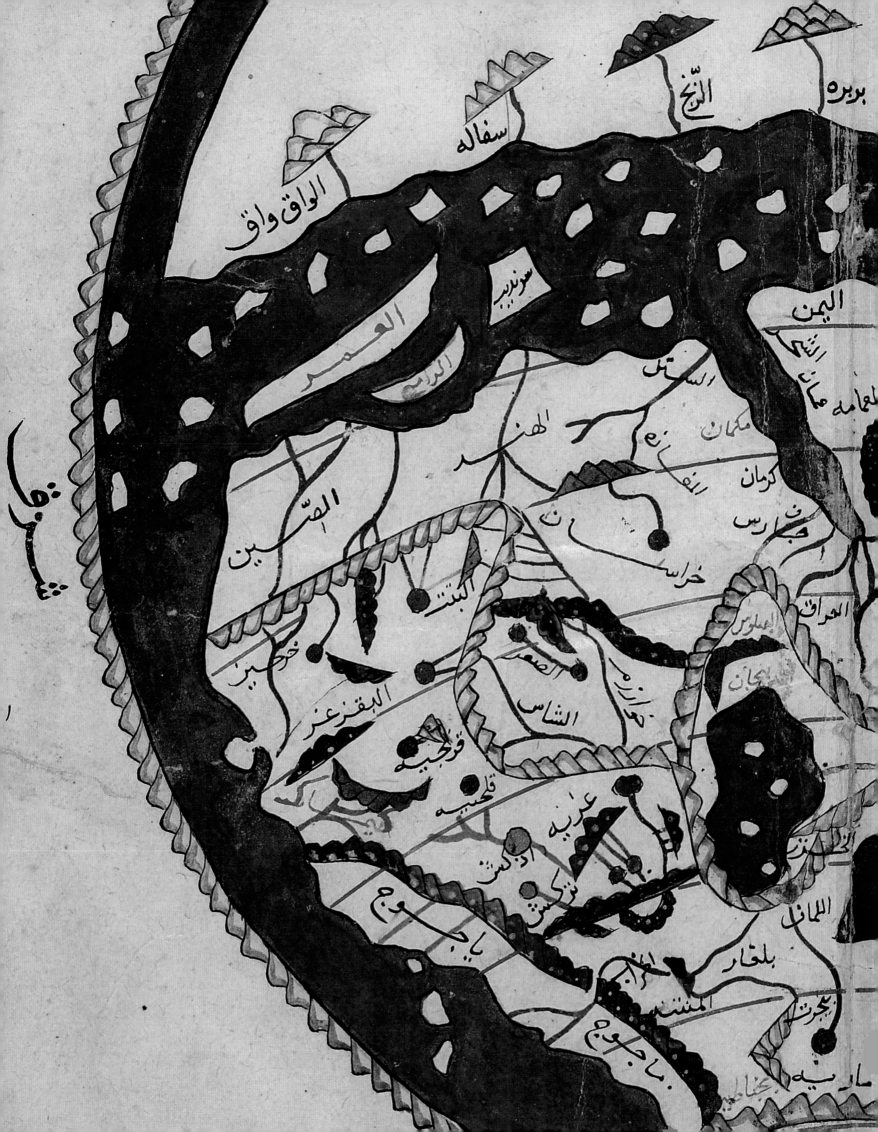

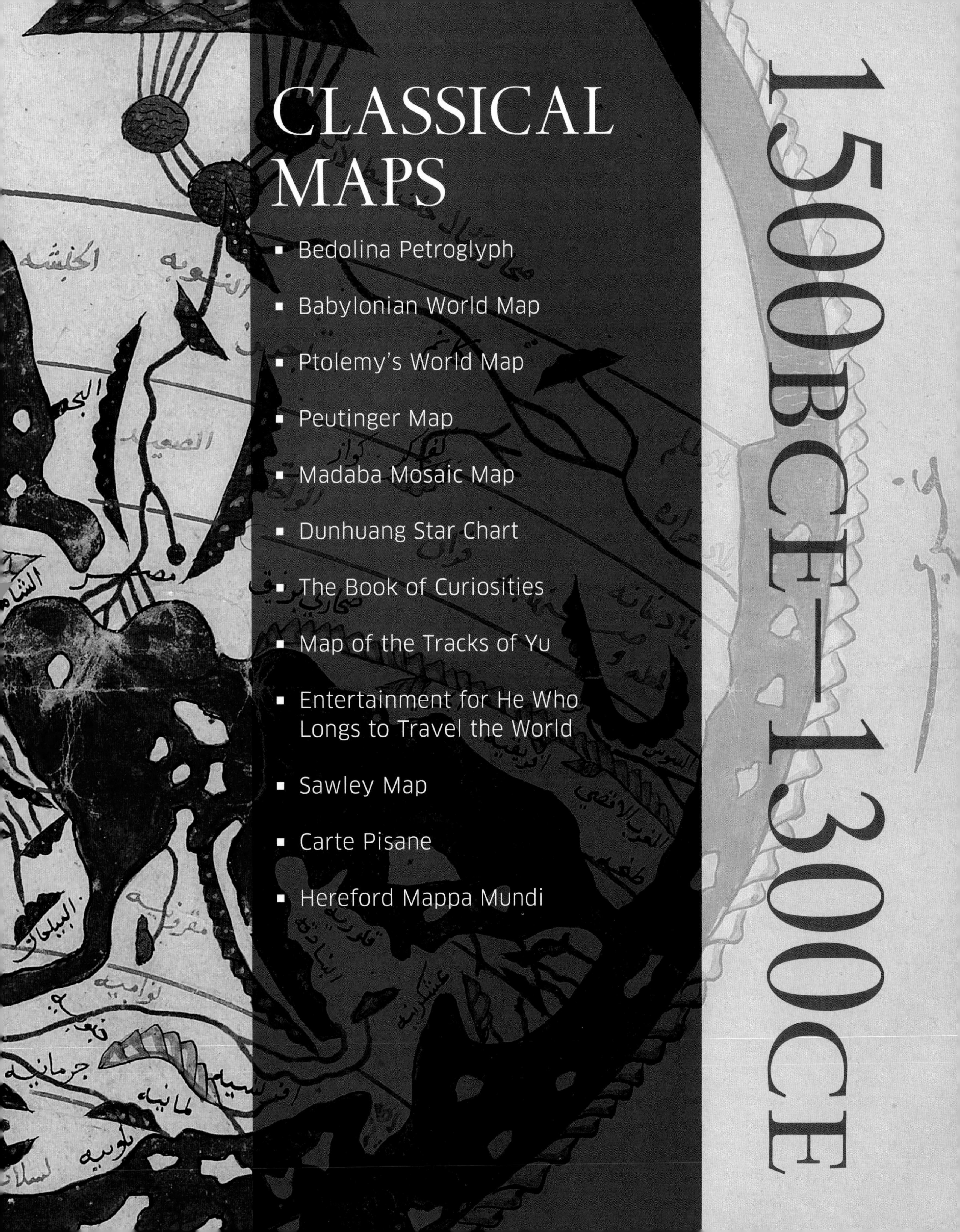

CLASSICAL MAPS

- Bedolina Petroglyph

- Babylonian World Map

- Ptolemy's World Map

- Peutinger Map

- Madaba Mosaic Map

- Dunhuang Star Chart

- The Book of Curiosities

- Map of the Tracks of Yu

- Entertainment for He Who Longs to Travel the World

- Sawley Map

- Carte Pisane

- Hereford Mappa Mundi

1500 BCE — 1300 CE

Bedolina Petroglyph

C.1500 BCE ▪ CARVED STONE ▪ 2.3M × 4.6M (7FT 6½IN × 15FT 1IN) ▪ CAPO DI PONTE, VALCAMONICA, ITALY

UNKNOWN

SCALE

Man's mapping impulse was manifested in the art of prehistoric man

CATHERINE DELANO SMITH, *THE HISTORY OF CARTOGRAPHY*

The urge to map the physical environment is a foundational human activity, and is probably related to mankind's acquisition of language. Some of the earliest maps can be dated as far back as the Upper Palaeolithic Period (40,000–10,000 BCE), when prehistoric cartographers carved their maps onto rocks. Known as petroglyphs, these were mostly rudimentary representations of local landscapes. One of the largest collections of petroglyphs can be found in Valcamonica, a valley in the Italian Alps.

Iron Age art

Of more than 200,000 incisions in Valcamonica, those discovered in the early 20th century at Bedolina in the Capo di Ponte region are some of the most interesting. At first, the *Bedolina Petroglyph* appeared to show simply the territory in the valley below, including the Oglio River. However, it is now believed that the network of geometric squares containing dots, interconnecting lines, and scenes of livestock and settlements are a more complex mix of sacred and topographical signs. Carved by members of an early Iron Age agricultural community called the Cammuni, they depict the beginning of the settlement and the hierarchical division of the region in the first millennium BCE. The map is, in fact, a cosmology – a partly imagined and highly abstract idealization of a settled landscape, as the Cammuni envisaged it.

ON **SITE**

Petroglyphs (the term is a combination of the Greek words for "stone" and "carve") date from more than 10,000 years ago. They were made by various methods of incising, picking, or carving an image into the rock's surface. The *Bedolina Petroglyph* was made using flint, quartz, or, later, metal tools to beat or peck (scratch) lines and shapes.

▲ **The stone map carvings** can still be seen today in the Alpine valley at Valcamonica.

Visual tour

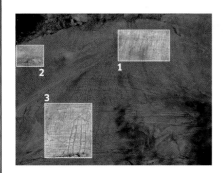

KEY

▶ **LIVESTOCK** The inclusion of livestock suggests that the map represents a crucial moment in human development, from a hunter-gatherer to a more agrarian culture. The anxieties over the community's food supply could have influenced these idealized carvings, a form of prayer to ensure agricultural success in the face of economic uncertainty.

▶ **HOUSES** These two wooden house structures with sloping roofs are more realistic but also cruder and larger in scale than the incisions higher up the rock. They appear to be much later additions, carved into the rock during the later Iron Age period, using metal tools. They do not seem to be part of the main composition.

▲ **FIELDS AND FIGURES** The dotted squares appear to be fields, suggesting an emerging land tenure system. Archaeologists speculate that each dot was pecked in as a votive offering to the dead or the gods to ensure successful harvests. Elsewhere warriors, animals, and even a ladder (perhaps into another world), have been added over subsequent centuries.

Babylonian World Map

C.750–C.500 BCE ▪ CLAY ▪ 12CM × 8CM (4¾IN × 3¼IN) ▪ BRITISH MUSEUM, LONDON, UK

UNKNOWN

SCALE

This broken and fragmented clay tablet is the earliest known map of the whole world. It was discovered near Sippar, southern Iraq, in 1882, although its maker and purpose are unknown. Despite its unprepossessing appearance, it offers one of the few remaining opportunities to glimpse the world view of the ancient Babylonian civilization. The top third of the tablet contains cuneiform (wedge-shaped) text, while the map occupies the bottom two-thirds. The world is depicted as a flat disc, surrounded by an encircling *marratu* (salt sea), in accordance with both the Babylonians' and the ancient Greeks' beliefs.

In Mesopotamian myth, the Sumerian king Etana was carried to heaven on an eagle and looked down on the Earth. This tablet imagines the world from above in a similar way and is as much an encapsulation of Babylonian mythology as it is a visualization of world geography. The map is composed around a central compass hole, placed in the middle of the Babylonian empire, and everything emanates from this sacred core. However, Babylonian power diminishes away from the centre and although many of the triangular zones around the map's edge have been lost, those that have survived are described in the cuneiform text as barbaric places beyond the limits of Babylonian civilization.

IN **CONTEXT**

In 1881, the Iraqi-born archaeologist Hormuzd Rassam discovered what became known as the *Babylonian World Map* while excavating a site near the ancient Babylonian city of Sippar. Rassam was looking for evidence of the biblical Flood, and because he could not read cuneiform text, he dismissed the tablet as of little importance. In fact, he did not even realize it was a map. It was only in the late 20th century that cuneiform scholars at the British Museum deciphered the tablet's text and discovered its significance.

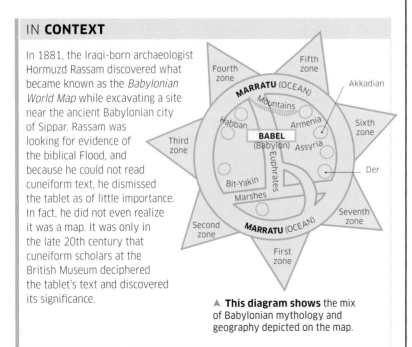

▲ **This diagram shows** the mix of Babylonian mythology and geography depicted on the map.

Visual tour

KEY

▶ **BABYLON AND THE EUPHRATES**
The vertical rectangle running down the middle of the map represents the Euphrates River, which empties into oblong swamps, beneath which is the ancient city of Susa. The upper horizontal rectangle shows Babylon, with surrounding cities and mountain ranges represented by circles, oblongs, and curves.

1

2

▲ **CUNEIFORM TEXT** Cuneiform is one of the earliest recorded systems of writing; on the top and back of the tablet, it describes Babylonian cosmology, including how the Earth was created "on top of the restless sea". The clay tablet is a physical manifestation of Babylonian accomplishments.

◀ **TRIANGLES** Labelled *nagû* (regions or provinces), these triangular zones represent the limits of the Babylonian world, describing dangerous places "where the sun is not seen", and terrifying beasts such as chameleons and lions. These are *terra incognita*, or unknown lands, common to many maps, and exerting an enduring repulsion and fascination.

3

Cuneiform
descriptions of
Babylonian
world view

Babylon

Marratu
(salt sea)

Euphrates
River

Ptolemy's World Map

c.150 CE ▪ VELLUM ▪ 57CM × 83CM (1FT 10½IN × 2FT 8¾IN) ▪ VATICAN LIBRARY, ROME, ITALY

SCALE

CLAUDIUS PTOLEMY

Claudius Ptolemy was the first classical scholar to apply geometry and mathematics to the study of the Earth. He produced a textbook called *Geography* (c.150 CE) that laid down the scientific method for projecting the globe onto a flat piece of paper (or in Ptolemy's case, papyrus). This textbook defined the study of geography, explained how to draw regional and world maps – using basic geometrical and mathematical principles to create two main projections – and listed 8,000 places within the classical world. Ptolemy's projection remained the template for geographers and mapmakers throughout the next millennia.

Mapping Ptolemy's world

Ptolemy introduced a basic graticule (a grid of coordinates), using latitude and longitude derived from centuries of Greek, Persian, Roman, and Arabic data, and then plotted the inhabited world, which he called the *ecumene*. Ptolemy's world picture stretched from the Canary Islands in the west to modern Korea in the east. His northernmost point was Thule – variously identified as somewhere in Scandinavia or the Orkney Islands (off Scotland) – with the south ending in Saharan Africa, which is joined to southeast Asia. There are no Americas or Pacific in Ptolemy's geography, which also overestimates the size of the Mediterranean, but underestimates the Earth's circumference. The *Geography*'s earliest surviving editions date from late 13th-century Byzantium, leading scholars to question whether Ptolemy ever actually produced a map himself.

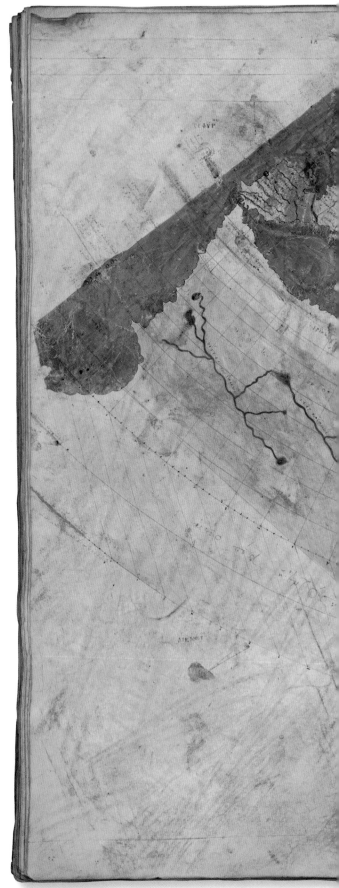

CLAUDIUS **PTOLEMY**

c.90–c.168 CE

Ptolemy was a Greco-Roman scholar who wrote some of the greatest scientific texts of the classical era during his time at the Royal Library of Alexandria in Egypt.

Ptolemy's surname suggests he was a native of Egypt – where the Ptolemaic dynasty had recently been overthrown by the Romans – and that he had some Greek ancestry; his forename indicates Roman citizenship. He worked at the Royal Library of Alexandria, the greatest library in the ancient world. It was built to gather together all known knowledge, and provided Ptolemy with the perfect location to develop a wide range of interests. As well as the *Geography*, he wrote books on music, optics, a highly influential study of geocentric astronomy, called the *Almagest*, and the definitive work on astrology, entitled the *Tetrabiblos*.

Geography is an imitation through drawing of the entire known part of the world

PTOLEMY, *GEOGRAPHY*

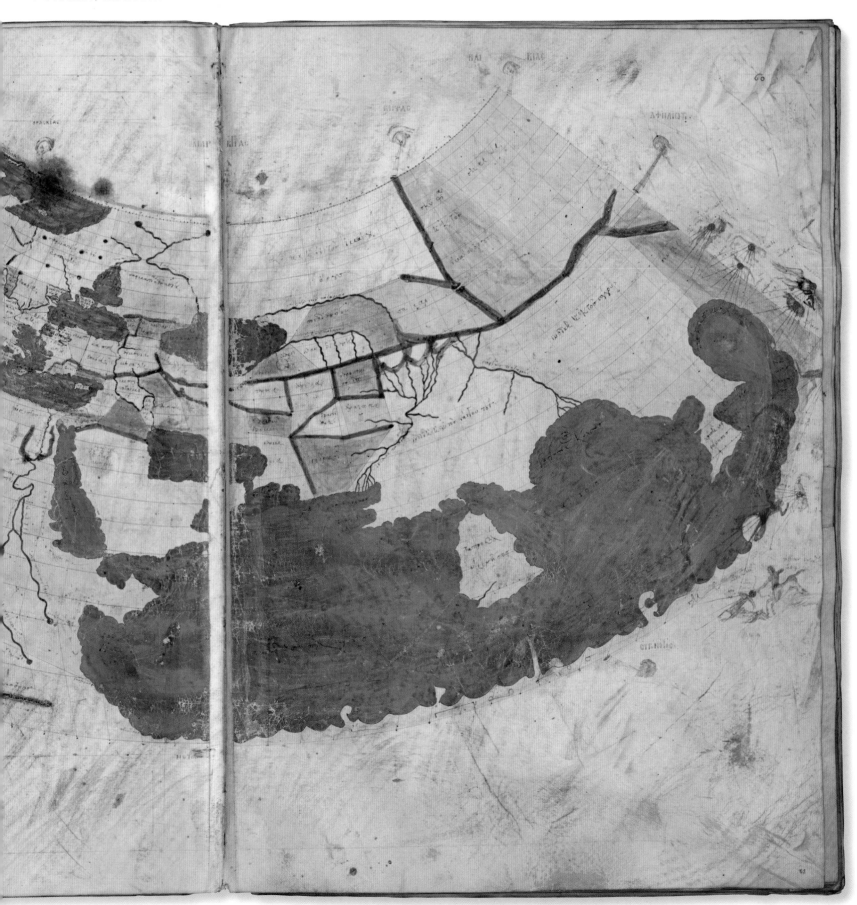

Visual tour

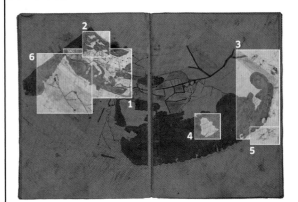

KEY

▶ **MEDITERRANEAN** For the Greeks, the *ecumene* (inhabited world) was centred on the Mediterranean. It is shown here with some accuracy, based on millennia of Greek and Roman knowledge. The North African coast is well mapped, including Ptolemy's home, Alexandria, as well as Byzantium, the Black Sea and, to a lesser extent, the Caspian Sea. Ptolemy overestimated the length of the Mediterranean at 61 degrees (its correct longitudinal length is 41 degrees), but this error was not corrected by western mapmakers for more than a thousand years.

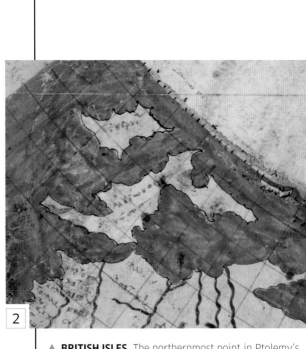

▲ **BRITISH ISLES** The northernmost point in Ptolemy's calculations is Thule, 63 degrees north, on the furthest edge of the British Isles. His regional description of the islands shows a reasonable knowledge of coasts, estuaries, and rivers, probably gleaned from Roman sources. These include Portsmouth, and the Rivers Trent, Humber, and Thames. In Scotland, he shows the River Clyde, while in Ireland, he includes the River Boyne. Despite Scotland being misaligned east to west (a common feature of late classical maps), it shows the outline of England quite well, and it is correctly aligned with Ireland.

▶ **THE LIMITS OF PTOLEMY'S WORLD** In the Far East, at the furthermost point on Ptolemy's map, is the enigmatic port of "Cattigara", situated 177 degrees east of the Canary Islands. Variously believed to be an important port in China, Korea, or even the west coast of the Americas, it is shown as part of a landmass running all the way to sub-Saharan Africa. To the west is the promontory known as "the Golden Chersonesus", believed to be the Malay Peninsula. Both places fascinated later explorers, including Columbus and da Gama.

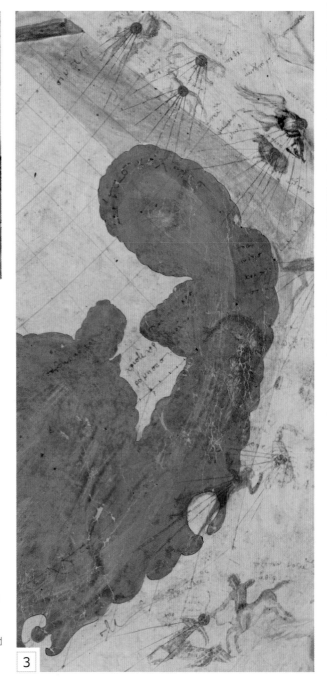

▲ **TAPROBANA** To the east, Ptolemy's geography becomes increasingly speculative. He includes the large island of "Taprobana", which is bigger than Britain and located somewhere near modern-day Sri Lanka. It is said to be twenty days sailing from India, whose coast is shown as almost entirely flat. The island is traversed by the Equator, and it could have been confused with various islands from eastern Africa to Indonesia.

ON **TECHNIQUE**

Ptolemy's greatest contribution to mapmaking was to propose two projections for mapping a spherical Earth onto a flat surface. The first was a conical projection, using a pin, string, and marker to draw curved parallels and a ruler to make straight meridians, which gave the impression of a cone. The second, more sophisticated projection, introduced curved parallels and meridians (see below). Ptolemy acknowledged that while both projections were imperfect, they provided the most effective solutions to a classical cartographic conundrum.

▲ **Ptolemy's curved projection** represented the Earth's appearance from space, and influenced generations of mapmakers.

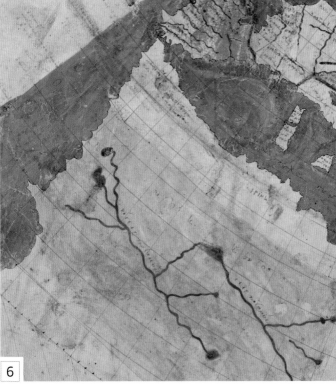

5

▲ **ZODIACAL SIGNS** Ptolemy was not only a geographer, but also an astrologer, and author of the *Tetrabiblos*, one of the most influential classical studies of how celestial events were believed to influence human affairs. Here his astrological observations are mapped onto the Earth, producing figures such as the archer Sagittarius, who in Greek myth was placed in the heavens to guide Jason and the Argonauts through Colchis in modern Georgia, but here is depicted in the Far East.

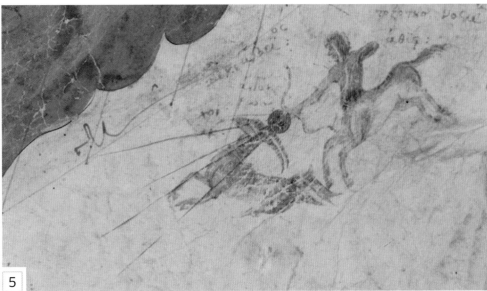

6

▲ **TERRA INCOGNITA** Ptolemy believed the inhabited world ended south of the Equator, along a parallel running through southern Egypt and Libya. His North Africa is connected by a series of lakes and rivers stretching westwards from the Fayum Oasis in Egypt, but everything 16 degrees south of the equator is simply unknown, or terra incognita.

Peutinger Map

c.300 CE ▪ PARCHMENT ▪ 30CM × 6.75M (1FT × 22FT 1¾IN)
▪ ÖSTERREICHISCHE NATIONALBIBLIOTHEK, VIENNA, AUSTRIA

UNKNOWN

SCALE

One of the most important maps in the history of cartography, the *Peutinger Map* provides a unique insight into Roman mapmaking. However, it is also an enigma: it is a copy, made in southern Germany around 1200, of a lost Roman original made around 300 CE. It is named after Konrad Peutinger, a German lawyer who inherited it from another scholar.

The most remarkable aspects of this map are its size and shape. It was created on parchment and cut into eleven sections, of which the fifth section, featuring Rome, and the eighth section, representing the eastern Mediterranean, are shown here. Nearly 7m (23ft) long and just 30cm (1ft) wide, it was probably carried round

in a Roman *capsa* (roll box). The map depicts the Roman Empire around 300 CE, stretching from the British Isles in the west to India in the east. Because shapes and distances running from east to west are on a much larger scale than those running from north to south, much of the topography seems strangely compressed, so that Rome seems just across the water from Carthage, and the Italian peninsula covers nearly three of the 11 sheets.

All roads lead to Rome

More than 100,000km (62,000 miles) of Roman roads are depicted, drawn in red, with distances between towns and cities recorded in miles, leagues, and even

A map without close match in **any period** or **culture worldwide**

RICHARD A. TALBERT, BRITISH-AMERICAN ANCIENT HISTORIAN AND CLASSICIST

Persian *parasangs*. Based on information taken from the *cursus publicus*, the official Roman transport system, the map depicts settlements, staging posts, spas, rivers, temples, and forests, and was once thought to be a road map for planning journeys or military campaigns. However, it was designed during the Tetrarchic period, when Rome was ruled by four emperors struggling to keep the fragmenting empire together. This suggests it might have been designed to convince Romans that their vast, harmonious, interconnected empire was still a reality, rather than a thing of the past.

KONRAD **PEUTINGER**

c.1465–c.1547

A diplomat, politician, and economist, Konrad Peutinger also achieved fame as an antiquarian, and amassed one of the largest libraries in northern Europe.

Widely regarded as one of the greatest humanist scholars of the European Renaissance, Konrad Peutinger was born in Augsburg in Germany (see pp.96–99). He studied law and classics at the renowned Italian universities of Bologna and Padua before returning to work as town clerk in Augsburg, and act as advisor to the Holy Roman Emperor, Maximilian I. He also began to amass a remarkable collection of ancient antiquities, in line with the Renaissance interest in the rebirth of classical culture, and became a famed book-hunter and collector of ancient manuscripts. In 1508, he inherited the extraordinary map that would eventually take his name.

Visual tour

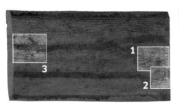

KEY

▶ **ROME** On this map, all roads really do lead to Rome, its symbolic heart. Rome is personified as an enthroned woman holding a globe, a shield, and a mitre, with 12 roads (all named) radiating outwards from the centre. The River Tiber is shown in green. On the left, the Via Triumphalis leads to St Peter's church (the word "Petrum" can be seen in red). This is a fascinating glimpse into the physical and symbolic perception of Rome.

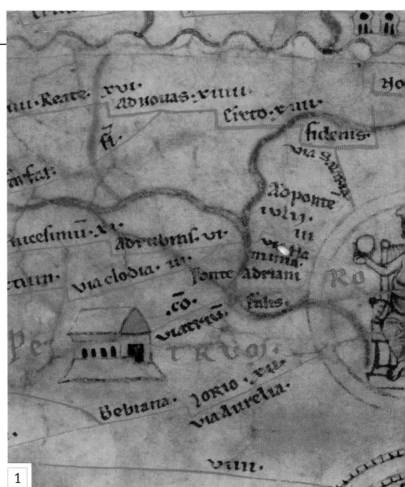

1

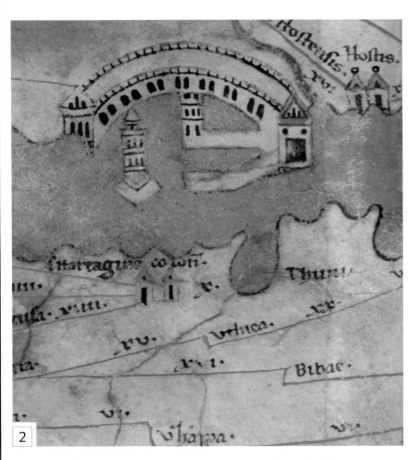

2

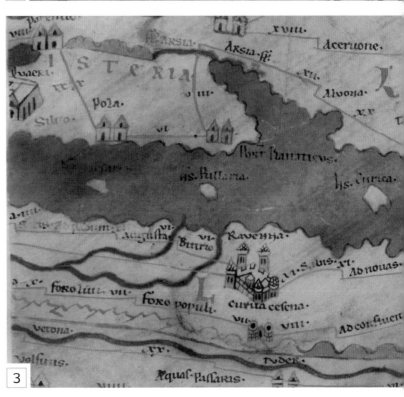

3

▲ **THE ROMAN HARBOUR AT OSTIA** Rome's ancient harbour of Ostia is depicted on a large scale in relief, although because of the wildly varying scales and north-to-south compression, it appears to be only a short distance across the water from the north African city of "Carthagine" (Carthage), the Roman Republic's great imperial rival, which appears to be a tiny settlement in comparison. The actual distance between Rome and Carthage is nearly 600km (370 miles), but the map is concerned with close imperial interests between the cities, not physical distance.

▲ **RAVENNA AND ISTRIA** Prominent walled cities such as Ravenna, near the Adriatic Sea, are shown in some detail, while smaller settlements are depicted as two-gabled buildings, with spas (top left) drawn as square buildings with courtyards. Mountains are indicated by brown wavy lines and the red roads show calculated distances between locations, estimated in Roman numerals above them. The map contains 2,700 places and distances, and 52 spas. Here the Adriatic Sea is so compressed that the land of "Istria" (mainly in modern-day Croatia) appears directly adjacent to the Italian coast.

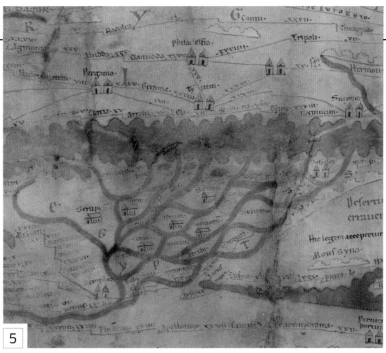

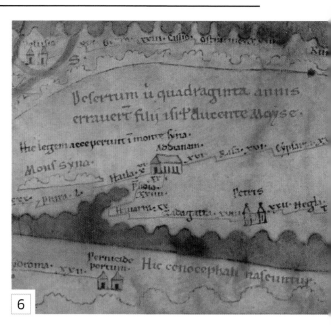

5

▲ **EGYPT, THE NILE, AND PERGAMON** The map's north-to-south compression is even more striking in its representation of Egypt. The Nile Delta is shown in all its sinuous detail, while the green strip of water running across the middle is the Mediterranean, squashed to accommodate the map's tiny breadth. The coastline running above is modern-day Turkey, dominated by the great ancient Greek city of Pergamon. The former Hellenic capital of "Pharos" (Alexandria) is shown very faintly in the bottom left. The distance between the two cities is in reality about 600km (370 miles).

6

▲ **MOUNT SINAI** The map shows an early Christian influence, which was very much part of Roman society around the time the map was made. On Mount Sinai, an inscription translates as: "the desert where the children of Israel wandered for forty years guided by Moses". The Red Sea runs across the middle, with an inscription at the bottom describing "cenocephali", supposedly a monstrous dog-headed people from northeast Africa.

4

▲ **CONSTANTINOPLE** Between 324 and 330 CE, the Roman emperor Constantine (272–337 CE), a recent convert to Christianity, founded Constantinople (modern-day Istanbul), establishing a new eastern part of the Roman empire. The new city appears almost as prominently as Rome, and Constantinople's female personification is strikingly similar to that of Rome's. She points to a column surmounted by the statue of a warrior, holding an orb and spear, which presumably depicts the city's founder, Constantine. The map captures the Roman Empire at a crucial moment, about to embrace Christianity and split into eastern and western halves.

IN **CONTEXT**

The *Peutinger Map* has traditionally been seen as encapsulating the Romans' more practical and logistical approach to mapping, especially in contrast to the scientific and often philosophical approach to maps developed by their Greek forbears, such as Ptolemy (see pp.24–27). It was believed the Romans adapted maps for use in land management, engineering, and military planning. However, recent research has identified more decorative, philosophical, and political aspects to Roman maps such as Peutinger's.

▲ **This *Orbis Terrarum* is a copy** of one of the Roman military maps commissioned by General Marcus Vipsanius Agrippa in about 20 CE. None of the original maps survived.

Madaba Mosaic Map

c.560 CE ▪ MOSAIC TILES ▪ 5M × 10.5M (16FT 5IN × 34FT 6IN) ▪ ST GEORGE'S CHURCH, MADABA, JORDAN

SCALE

UNKNOWN

The finest surviving example of Byzantine mapmaking and the oldest known floor map made from tiles, the Madaba mosaic map can be seen today on the floor of St George's church at Madaba in central Jordan. Only a quarter of the original mosaic survives, but what remains is one of the greatest early maps of the Holy Land, from the Jordan Valley to the Canopic branch of the Nile. It is oriented with north at the top and Jerusalem at the centre, in keeping with early Christian maps of the region.

A map for the faithful

This map depicts topographical elements with great detail; over 150 places are named in Byzantine Greek. The map's scale varies between 1:15,000 for Judaea and 1:1,600 for Jerusalem – the greater level of detail afforded to the latter is probably due to its importance. It is also packed with vivid scenes of fishing boats, bridges, lions, gazelles, fish, architectural features, and passages from the Bible. Many of these passages are taken from the Roman Christian scholar Eusebius of Caesarea's *Onomasticon* (c.320 CE), a book of biblical place names. This suggests that the map was intended to instruct the faithful and those on religious pilgrimage.

The Jordan river is shown filled with fish

Wild animals decorate the blank spaces of the mosaic

The map is labelled in Byzantine Greek

ON **SITE**

▲ **St George's church, Madaba** is still a popular destination for pilgrims.

The town of Madaba was mostly destroyed in an earthquake in 746 CE, and the mosaic lay hidden until 1884, when it was uncovered during construction of St George's, a new Greek Orthodox church built on the site of another ancient building. The church was completed with the mosaic located in its apse, where the surviving parts of the map can still be seen today. Unfortunately the mosaic suffered various degrees of damage between its rediscovery and the 1960s, when work was undertaken to conserve and restore its surviving parts. Copies of the map are also on display in Germany and Jerusalem.

A snapshot of the then contemporary world in all its wondrous tolerance and coherence

DENIS WOOD, *RETHINKING THE POWER OF MAPS*

Visual tour

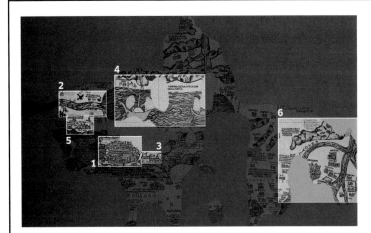

KEY

▼ **JORDAN RIVER** The mighty river where Jesus was baptized and where the Israelites crossed into the Promised Land is shown teeming with fish and flanked by lions chasing gazelle. Pulley ferries (below left) are also shown. Visible to the top right is "Gilgal", where Joshua planted 12 memorial stones, while to the lower right is "Bethabara", the church and baptismal site of St John the Baptist.

▼ **JERUSALEM** At the centre of the map is the walled "Holy City of Jerusalem", the heart of Christianity, rendered in astonishing detail. This includes 21 towers and a variety of aerial perspectives. To the left (or north) is the Damascus Gate, leading to the red-tiled Basilica on Mount Sion. The Church of the Holy Sepulchre is shown in the central foreground.

▶ **BETHLEHEM** The map's religious topography is so precise that even the individual buildings in Bethlehem associated with Christ's birth are rendered in minute detail, based on scrupulous reference to the sources of the time. It clearly depicts the Basilica of the Nativity connected to a red-tiled building, and a yellow dwelling with a tower. Below are the Mountain of Judah and "Ephratha", a location mentioned in Genesis, whose name means "fruitful".

▶ **BOATS ON THE DEAD SEA** "Salt, also Pitch Lake, also the Dead Sea" is how the mosaic's text describes this region. This sea was famed as a spa visited by King Herod, and the names refer to the bitumen (or pitch) extracted from it, and fact that it contains no living things. The features of the sailors in each boat, shown plying their trade across the river, have been erased by iconoclasts following the Byzantine bans on the depiction of the human form in art.

▲ **JERICHO** Described in the Old Testament as the "City of Palm Trees", Jericho is also the location of the Israelites' first battle after crossing the Jordan in their conquest of the Promised Land. It is a fertile, prosperous oasis, surrounded by nine palm trees and boasting four towers and three churches.

◀ **SINAI REGION** Here Egypt, the Nile, its tributaries, and the Sinai region are all shown. The rocky Sinai desert is at the top, full of references to the Israelites and the Exodus from Egypt into the Promised Land. At the top left is "Raphidim", where the Amalekites attacked the Israelites; below is "The Desert of Zin where we were sent down the manna", as described in the Book of Exodus.

ON **TECHNIQUE**

The map was laid using *tesserae*, small cube-shaped tiles of limestone, marble, or coloured glass common in Roman and Byzantine mosaics. However, this required enormous time, skill, and money. A master craftsman would have first sketched out the map's outlines before an artist laid down coloured tesserae for cities, towns, mountains, rivers, figures, and inscriptions.

▲ **This modern Christian** mosaic in Madaba was made using the same methods as the historic map.

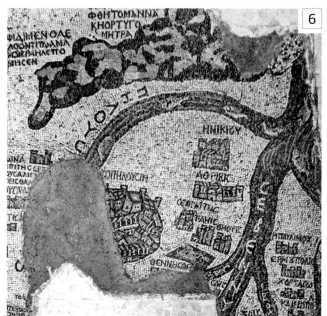

Dunhuang Star Chart

C.649–684 CE ▪ INK ON PAPER ▪ 24CM × 3.94M (9½IN × 12FT 11IN) ▪ BRITISH LIBRARY, LONDON, UK

SCALE

LI CHUNFENG

Many early civilizations observed the stars, often motivated by the belief that changes and activity in the heavens dictated various aspects of life on Earth. Of these civilizations, China was the most advanced in astronomy and celestial mapmaking. The *Dunhuang Star Chart* is the oldest known map of the heavens, made around 649–684 CE, during the Tang Dynasty. It was discovered by distinguished British-Hungarian archaeologist Aurel Stein in 1907, among 40,000 other precious manuscripts in the so-called "Caves of the Thousand Buddhas" in the town of Dunhuang, situated on the Silk Road in the northwest of China.

The chart is drawn on the finest paper, and is nearly 4m (13ft) in length but just 24cm (9½in) wide. Read from right to left, it begins with a text describing the art of uranomancy – divining events by observing natural phenomena in the sky – and shows a variety of "celestial vapours" (cloud formations). This is followed by 12 hour-angle star maps (each covering a twelfth of the sky), and a final star chart of the circumpolar regions.

An impressive undertaking

In total, these 13 maps display the entire sky visible from the northern hemisphere. An estimated 1,339 stars are shown, grouped into 257 constellations, all

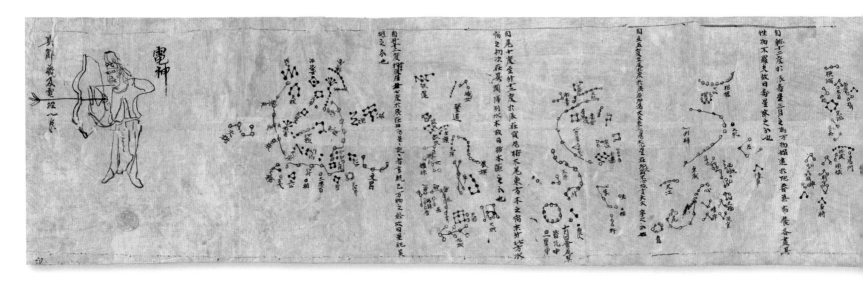

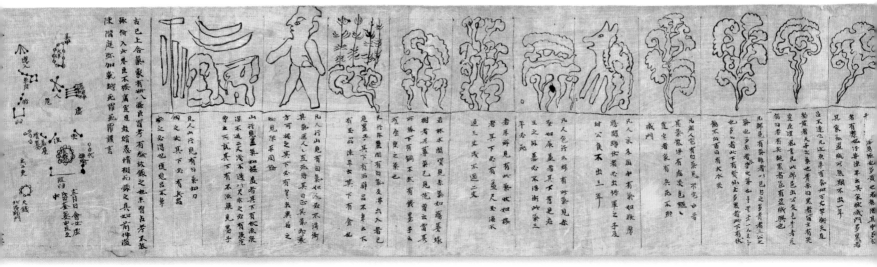

The oldest complete preserved star atlas known from any civilization

JEAN-MARC BONNET-BIDAUD, FRANCOISE PRADERIE, AND SUSAN WHITFIELD, *THE DUNHUANG CHINESE SKY*

apparently visible to the naked eye from the Chinese imperial observatory. The star chart's author, believed to be astronomer Li Chunfeng, drew on the work of three earlier astronomical texts cataloguing the heaven's stars and their astrological divinations, the earliest of which predates his star charts by more than 700 years. Some of the stars' colours refer to the astronomers who originally identified them, while many of the names and descriptions of the constellations offer a revealing insight into the preoccupations of everyday Chinese life, such as social status and personal fortune.

LI **CHUNFENG**

602–670 CE

The star chart's commentary mentions "your servant Li Chunfeng", which suggests that this eminent Taoist mathematician, astronomer and historian was involved in its creation.

Li was the official court astronomer, historian, and director of the Imperial Astronomy Bureau during the early Tang Dynasty. He was renowned for his reforming work on the Chinese calendar, which was introduced in 665 CE. Entitled the Linde calendar, it introduced an "intercalary month" – a concept similar to the Western leap year – every three years. This task was particularly important as it improved the ability to accurately predict eclipses, which were considered crucial for ruling imperial policy. Li also wrote imperial mathematical manuals and texts on astrology, meteorology, numerology, and music. These works were highly respected by his contemporaries, as were the Dunhuang charts.

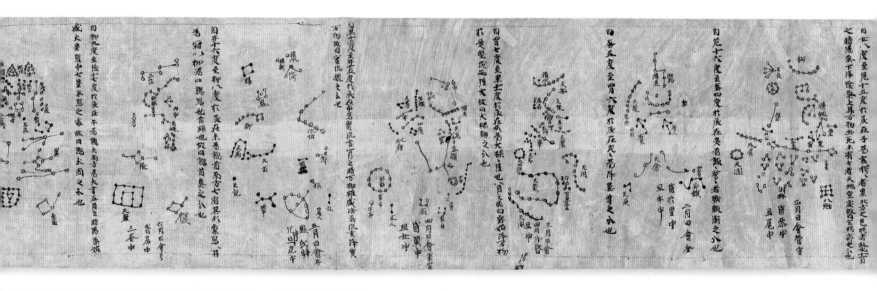

Visual tour

KEY

▶ **NORTH CIRCUMPOLAR REGION** The northern region of the sky is the most important for Chinese astronomy. The heavens reflected Chinese social hierarchy, and this area was considered the most significant, hence its position as the final map. Each asterism (star group) was identified with figures and locations in the imperial court. A total of 144 stars are shown, all revolving around the polar (pivot) star, the hazy red dot at the very centre, which represented the emperor. To its right are four stars representing the emperor's advisors.

▼ **THE BOWMAN** The chart ends with the image of a bowman wearing traditional clothes, firing an arrow. The caption to his right suggests that he is the Taoist deity of lightning – a phenomenon thought to be caused by the release of his arrow. He is an appropriate figure to conclude the manuscript, which begins and ends on the theme of the divination of meteorological events upon human fate.

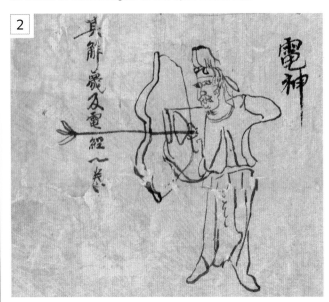

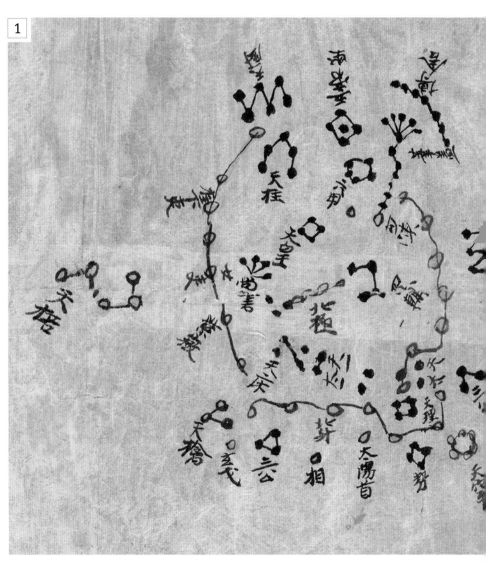

▶ **CLOUD DIVINATION** The chart begins with 26 delicate drawings of different cloud formations, along with 80 explanatory columns of text describing the art of uranomancy – all of which are deemed "reliable" because they have been "tested". A cloud "in the form of a leaping or crouching wolf", for example, is considered propitious and "the family will certainly bear a son who will become a general and will be conferred a rank of nobility". Other cloud formations are less fortuitous and presage death and destruction.

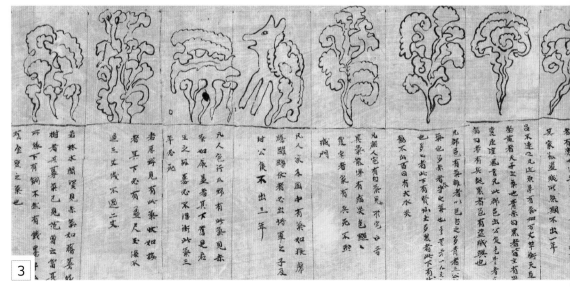

▼ **ORION CONSTELLATION** The constellation of Orion was among the oldest known to Chinese astronomers. It is one of the most visible constellations due to its brightness and location along the celestial equator. Here it is identified as "Shen", a great hunter. It is one of the few cases in which Chinese and Western astronomers agree on the shape of a constellation.

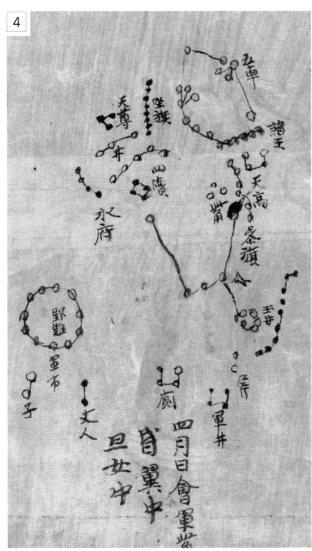

IN **CONTEXT**

The ancient Chinese believed that events on Earth were shaped by the heavens. The ability to predict events such as eclipses and comets and their terrestrial manifestation conferred great power on Chinese astronomers, who were required to produce calendars foretelling certain celestial changes. The sky was divided into regions or *gong* (palaces), and the four cardinal directions were each represented by an animal – a black tortoise, blue dragon, red bird, and white tiger. The key central region represented the imperial ruler, his throne, and advisors (the symbol for China means "the middle"). Also, 28 celestial segments (or mansions) contain the thousands of stars shown on celestial charts.

▲ **Ancient Chinese astronomers** were highly influential in government.

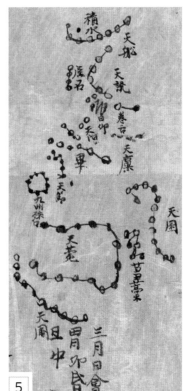

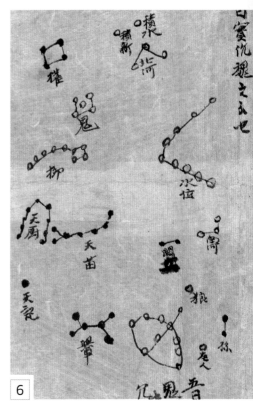

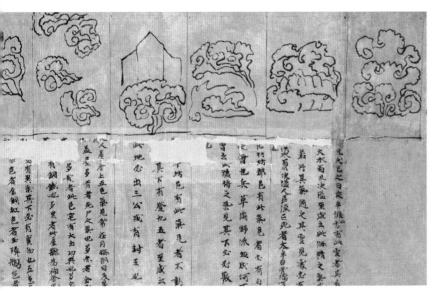

▲ **CANOPUS** The second brightest star, Canopus is located deep in the southern sky. In Greek myth, Canopus was a helmsman and the star was used as a southern pole star. The Chinese called it the "Old Man" and associated it with the god of longevity. On this chart, however, it is shown far too close to the Equator.

▲ **PLEIADES AND HYADES** The constellation of Taurus consists of two star groups, the Pleiades and Hyades, both shown here on the fourth chart. The Pleiades are known in the West as the Seven Sisters. They are located here in the celestial station known as "Da Liang" and described by the Chinese as "Mao", the hairy head of a white tiger.

The Book of Curiosities

c.1020–1050 ▪ INK ON PAPER ▪ 24CM × 32CM (9½IN × 12½IN) ▪ BODLEIAN LIBRARY, OXFORD, UK

UNKNOWN

SCALE

In 2002, the Bodleian Library in Oxford, UK, acquired an anonymous Arabic treatise compiled in Egypt between 1020 and 1050 during the Fatimid Dynasty, entitled *Kitāb Gharā'ib al-funūn wa-mulah al-'uyūn* ("The Book of Curiosities of the Sciences and Marvels for the Eyes"). The manuscript, which describes the heavens and the Earth according to Muslim astronomers, scholars, and travellers, has transformed our understanding of early Islamic cosmology and geography. It includes several maps of the inhabited world, including two world maps, one of which is circular, while the other, shown here, is rectangular. This map is unlike any other surviving map from either the Christian or Muslim medieval worlds.

A medieval Islamic perspective

The map is oriented with south at the top, which is typical of Islamic cartography of the time, with the Arabian Peninsula and Mecca displayed with particular prominence. Europe, to the lower right, is dominated by a huge Iberian peninsula, concentrated on Muslim-controlled Spain. Meanwhile, North Africa dwarfs Italy and Greece, and it is shown in far greater detail, particularly Egypt and the complicated tributaries that make up the source of the Nile. Arabia is dominated by Mecca and is depicted at twice the size of India and Persia, while Central Asia includes the Euphrates and Tigris rivers and the city of Constantinople (now Istanbul), which is noted as having a "Christian creed". This region also retains mythological elements, such as Alexander the Great's fabled wall built to keep out the monstrous Gog and Magog. Further east, India and China are shown, but in increasingly hazy detail, while the limits of the inhabited world are represented on the far left by the mysterious "Island of the Jewel". The scale bar along the top suggests the use of mathematical applications hitherto unknown in medieval mapmaking. Both book and map are heavily indebted to Greek sources, particularly the works of Ptolemy (see pp.24–27), as well as a variety of Arabic and Islamic authorities.

> God has divided Earth into regions, and made some regions higher and others lower; and He made the constitution of the inhabitants of each region to correspond with the nature of the region

THE BOOK OF CURIOSITIES

Visual tour

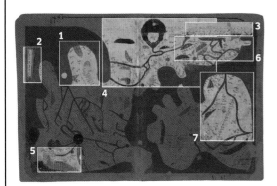

KEY

▶ **ARABIAN PENINSULA AND MECCA** The holy Muslim city of Mecca (centre right) dominates the Arabian peninsula. Unlike other cities on the map, it is symbolized by a horseshoe shape, which may refer to the *Hatim*, a semi-circular wall opposite the *Kaaba*, the city's holiest building. The sacred geography of early Islam is emphasized, including the cities of Medina, Sana'a, and Muscat, indicated by red dots. Yemen's Hadramawt mountain range (top) is shown coloured red.

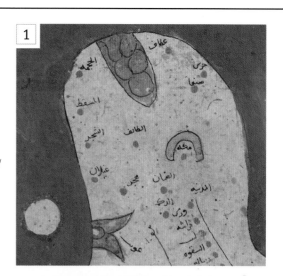

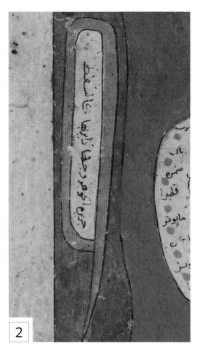

▲ **ISLAND OF THE JEWEL** The location of this enigmatic island, placed at the easternmost limits of the inhabited world, is taken from the works of the Persian mathematician and geographer al-Khwārizmī (c.780–c.847 CE). He describes the island as situated close to the Equator and near the "Sea of Darkness" – the Atlantic. Here it lies east of India and China; it may be what is now Taiwan, although its true location and identity remain a mystery.

▶ **SCALE BAR** The scale bar is one of the earliest recorded on any map. It increases from the right in units of five degrees, suggesting a sophisticated attempt to measure the Earth mathematically. The numbering stops on the left, suggesting the mapmaker tried unsuccessfully to copy an earlier, more technical map.

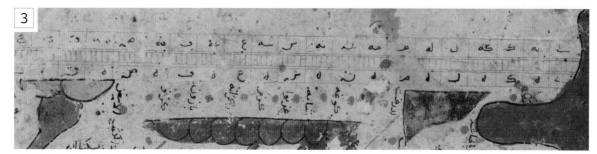

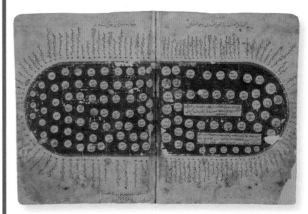

Like many Islamic mapmakers, the author of *The Book of Curiosities* divides the world up into seven climes (taken from the Greek *klimata*, meaning "incline"). Aristotle believed that climate influenced the degree and nature of the Earth's inhabited regions. They included the "intemperate", equatorial regions, the frozen wastelands of the north, and the "temperate" areas of the Mediterranean. Islamic scholars embraced the concept and it appears in *The Book of Curiosities* in the description of climes stretching from the first in the south, which runs through Africa ("Land of the Scorching Heat"), India, and China, to the seventh in the far north, which runs through Scandinavia and includes descriptions of an island populated solely by women. The fourth clime runs through Rhodes and Babylon, and "has the best constitution and disposition".

▲ **Shown here in** *The Book of Curiosities*, the Mediterranean region was considered "temperate" in the Aristotelian model.

▲ **ALEXANDER'S BARRIER** Both Christianity and Islam believed in the story of Alexander the Great walling in the mythical monsters Gog and Magog in the Caucasus Mountains. Here the wall is shown as a "barrier which the possessor of two horns built". Alexander was associated with ram's horns – a symbol of power and virility.

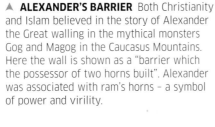

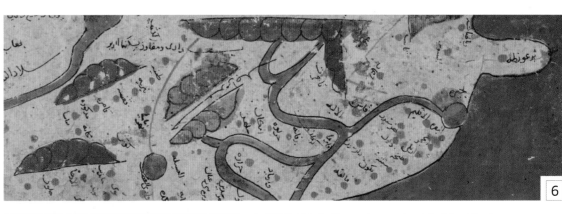

▲ **MOROCCO** As the westernmost point of the Islamic world, Morocco's rivers, mountains, holy cities, and commercial routes are shown in great detail – the cities of Tangier and Fez are particularly prominent. The struggle to impose Islamic authority on the region is also shown: in the top right is Barghwatah, a Berber confederation that ruled much of the coastal regions; inland are "deserts inhabited by the Berbers".

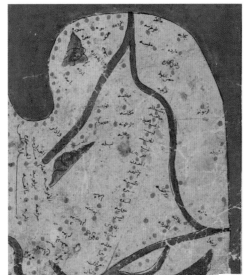

▲ **THE NILE AND THE "MOUNTAINS OF THE MOON"** The source of the Nile has fascinated explorers since the time of the ancient Greeks. Here it is represented according to the theories of the Greek philosopher Diogenes (c.404–323 BCE). He mistakenly believed it originated in a central African mountain range from which ten rivers flowed into lakes running into the Nile. The myth was only disproved in the 19th century.

◄ **ANDALUSIA** By the 11th century, Moorish Spain, or "Andalusia", was fragmenting politically because of factionalism among competing Muslim dynasties and the resurgence of Christianity. The region is described as being "20 days' journey in breadth", and shows a diagonal itinerary, outlined in red dots, running through Lisbon and Seville to Almeria, with Córdoba, the home of the Umayyad Caliphate, depicted just above the Guadalquivir river (top left).

Map of the Tracks of Yu

C.1136 ▪ STELE ▪ 84CM × 82CM (2FT 9IN × 2FT 8¼IN) ▪ BEILIN BOWUGUAN, SHAANXI, CHINA

SCALE

UNKNOWN

Possibly the most famous of all Chinese maps is known as the *Yu ji tu*, or "Map of the tracks of Yu" (*tu* is Chinese for "map"). It dates from the Song Dynasty (960–1279) and takes its name from the legendary exploits of the celebrated Chinese ruler Yu the Great, renowned as a mythical creator of the landscape and controller of floods. It is carved on a stone slab or stele and gives a remarkably accurate rendition of China's outline, particularly its rivers and coasts. It is also the first known map to use a cartographic grid showing scale. The sides of each of its 5,000 squares represent 100 Chinese *li* (just over 50km, or 30 miles), giving an estimated scale of 1:4,500,000. Despite its accuracy, the map gives erroneous origins for the Yellow River, showing the sources named by Yu the Great. It is a singularly Chinese fusion of cartographic accuracy and written legend.

> The most **remarkable** cartographic work of its age in **any culture**

JOSEPH NEEDHAM, BRITISH HISTORIAN

IN **CONTEXT**

Chinese rulers and leaders from the 7th century CE onwards used stelae, (monumental stone slabs) like this one, inscribed with text and pictures, to inform and instruct the public about events and to record religious and other inscriptions. Carving a map on to a stele using a square grid is an immensely skilled activity. The resulting image, however, allowed unskilled users to copy the map by making a rubbing. Such rubbings on paper circulated widely among many Chinese communities.

▲ **This stele** bears a map of the ancient city of Xian.

Visual tour

KEY

1

▲ **THE LEGEND** This explains that the map shows "provinces and prefectures" and "mountain and river names" taken "from past and present", acknowledging the mix of cartographic fact and myth.

2

◀ **THE YELLOW RIVER** Yu claimed that the region of Jishi, shown here, marked the river's origins. The true source, however, lay in the Kunlun mountain range.

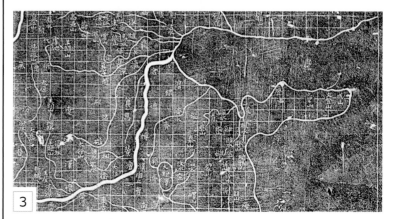

3

▲ **SHANDONG PENINSULA** One of the oldest and most populous regions of China, the peninsula is strategically important because it includes the lower reaches of the Yellow River. The map shows it in extraordinary detail.

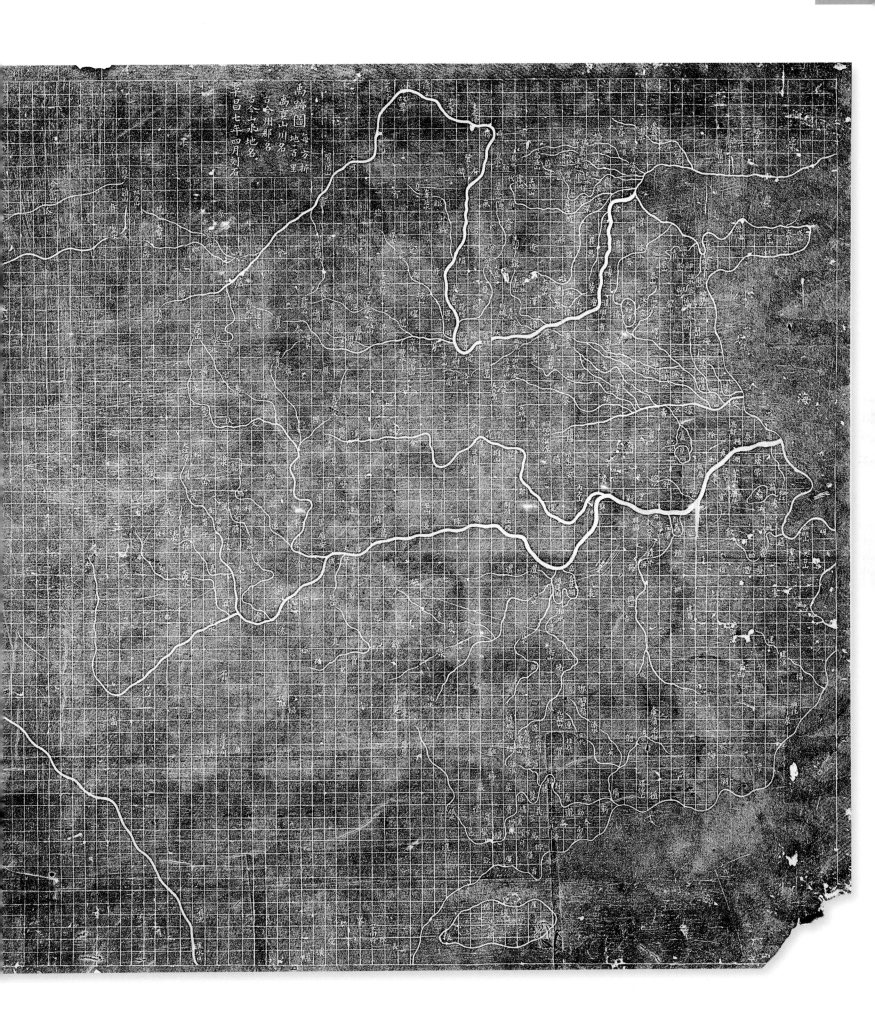

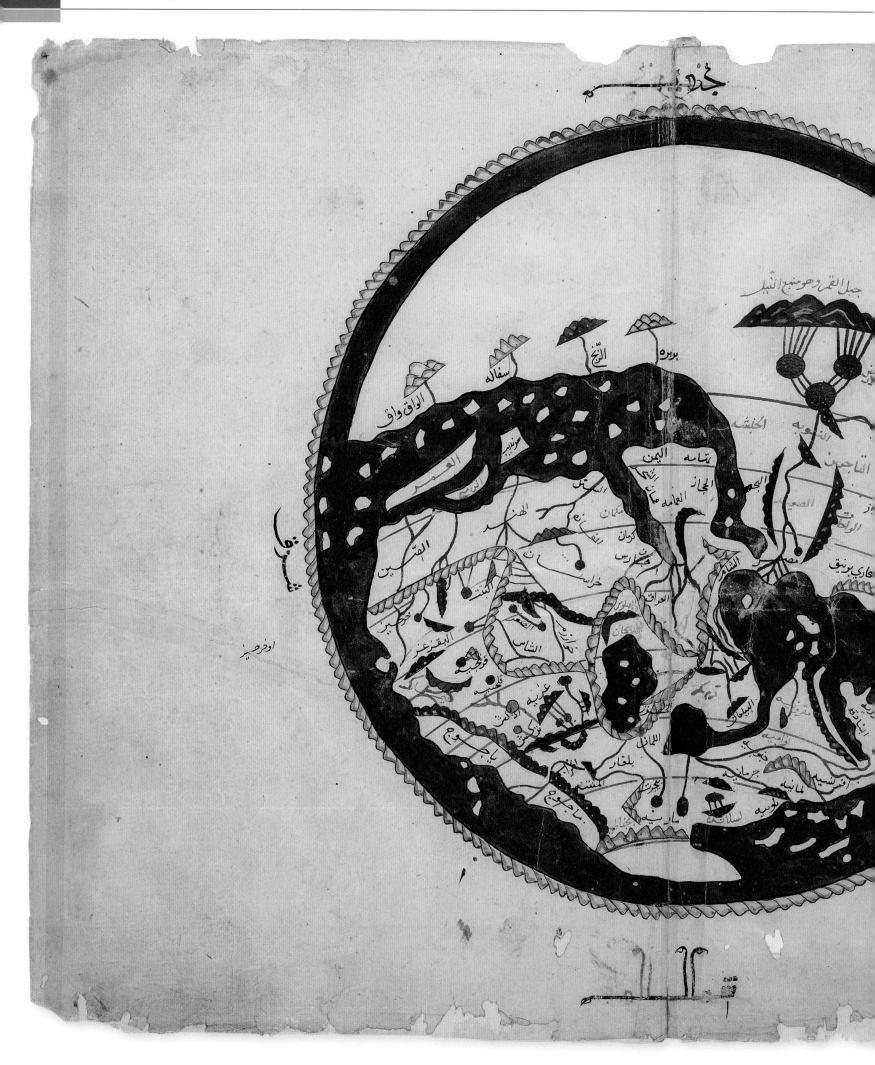

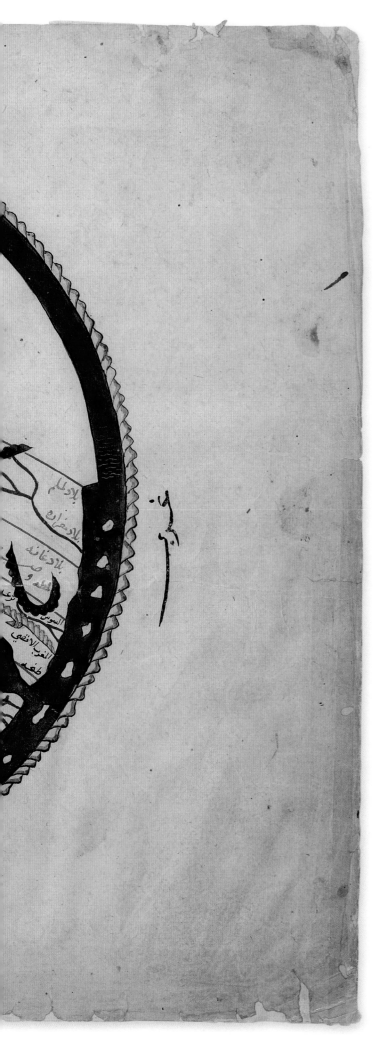

Entertainment for He Who Longs to Travel the World

1154 ▪ MANUSCRIPT ▪ 21CM × 30CM (8¼IN × 11¾IN)
▪ BODLEIAN LIBRARY, OXFORD, UK

SCALE

AL-SHARĪF AL-IDRĪSĪ

One of the most remarkable products of medieval mapmaking, this circular world map is taken from a geographical book, *Kitāb nuzhat al-mushtāq fī ikhtirāq al-āfāq* ("Entertainment for He Who Longs to Travel the World"), written by the Muslim Arab scholar al-Sharīf al-Idrīsī. Composed in Sicily under the patronage of the island's Norman ruler Roger II, al-Idrīsī's book contains 70 regional maps covering the inhabited world, and begins with this world map. Perhaps the most surprising feature is its orientation: south is at the top. Most early Islamic world maps were oriented this way because many of the communities that converted to Islam in the 7th and 8th centuries lived north of Mecca, and so faced south during prayers. The Earth is encircled by sea and surrounded by fire, a concept taken from the Qur'an. Al-Idrīsī's African homeland is vast, dominated by the mountains believed to be the source of the Nile; Europe is very sketchy (although with an unsurprisingly large Sicily, given its patron), and Arabia (including Mecca) occupies the central area.

AL-SHARĪF **AL-IDRĪSĪ**

c.1099-1161

Al-Sharīf al-Idrīsī was a geographer, traveller, Egyptologist, and mapmaker who produced some of the most accurate maps of his time. He studied at the great Muslim university in Córdoba, Spain, and is then reputed to have travelled across western Europe.

Few mapmakers can claim descent from a religious prophet, but al-Idrīsī, who was born in Ceuta on the North African coast, was a member of the Hammūdid family, which traced its ancestry back to Muhammad. In the late 1130s, al-Idrīsī arrived in Norman Sicily, where he spent the next three decades working for its wily, ambitious king, Roger II, and moving among its mixed intellectual heritage of Christian, Greek, Jewish, and Muslim scholarship. Al-Idrīsī published geographical and medical treatises, as well as making various maps and globes, before returning to North Africa.

Visual tour

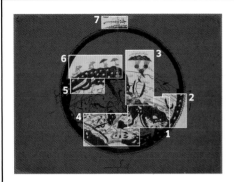

KEY

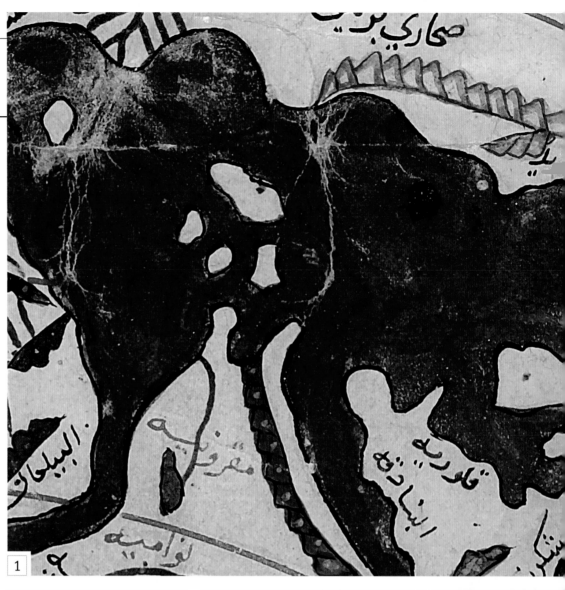

▶ **SICILY RULES** Right of the map's centre, the roughly triangular shape of Roger II's Kingdom of Sicily – al-Idrisi's adopted home – dominates the Mediterranean. Although it is the region's largest island, al-Idrīsī has magnified it so that it looks four times the size of Sardinia (below it), which is in reality only slightly smaller.

1

2

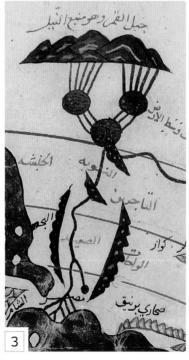

3

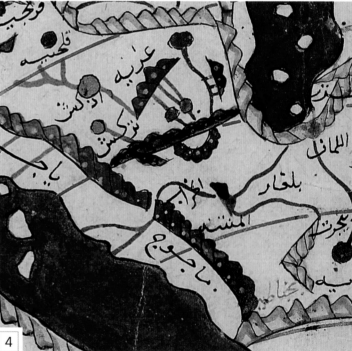

4

▲ **ISLANDS AT THE WORLD'S END** Along the map's edge, running from Iberia down the west coast of Africa, al-Idrīsī drew half a dozen rather amorphous islands. These are Ptolemy's Fortunate Isles, probably the Canaries, from where most classical mapmakers established their prime meridian. Al-Idrīsī called them "al-Khalidat", placing them in the Atlantic, known to Muslim mapmakers as the Sea of Shadows.

▲ **THE NILE'S ORIGIN** Like Ptolemy and many Muslim mapmakers, al-Idrīsī was fascinated by the Nile's source. The Greeks believed that the river originated in a range of snow-capped mountains and then flowed into a series of vast lakes, as shown here.

▲ **CENTRAL ASIA** Throughout this region, al-Idrīsī has tried to develop a cartographic vocabulary, using golden triangulated peaks to represent mountain ranges, browns and greys to show rivers and lakes, and black inscriptions describing cities, towns, and settlements. At the top, the Black Sea dominates, with a very oddly shaped eastern Aegean Sea to the right.

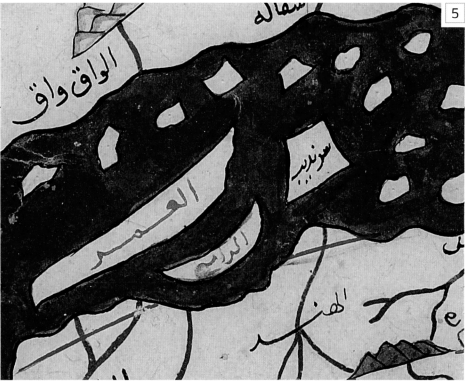

◀ **A DISTANT SEA: THE INDIAN OCEAN** East of the Arabian Peninsula, al-Idrīsī's knowledge ran out and the map dissolves into contradictory fragments drawn from myth, Greek history, and early Islamic geography. There is some acknowledgment of the region's plethora of islands and archipelagos, but India loses its distinctive peninsula altogether, and Sri Lanka (labelled "Taprobana") is shown as a casual smear to the left. Even Ptolemy's confident but erroneous grasp of the region's geography had disappeared by this point.

◀ **AN IMAGINARY SOUTHERN CONTINENT** The theory of *klimata* (see p.43) led the Greeks to believe in a temperate southern zone of the "antipodes", the Greek term for "those with the feet opposite", or people inhabiting the other side of the world. Al-Idrīsī's southern continent is schematic, an extension of a vast Africa, featuring mountain ranges.

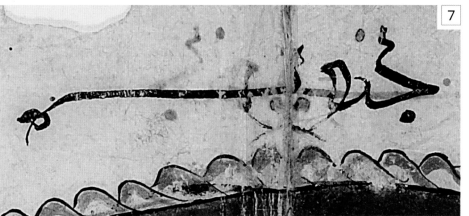

◀ **QUR'ANIC GEOGRAPHY** The map's four cardinal directions are labelled, and there is a short inscription drawn from the Qur'anic belief in the creation of seven "firmaments", including a disc-shaped Earth, encircled by water. The map contains no other theological beliefs or assumptions about earthly creation. There are no monsters, just a curious naturalism about the world's shape and extent.

IN **CONTEXT**

Al-Idrīsī drew on a rich tradition of Islamic mapmaking going back to the establishment of the Abbasid Caliphate in Baghdad at the end of the 8th century. Most early Islamic mapmakers borrowed from Greek geography, especially the work of Ptolemy (see pp.24–27), taking his concept of dividing the Earth into latitudes, called *klimata* (see p.43), and adapting it to describe Islamic *iqlim* (provinces). Unlike Christian *mappae mundi* (see pp.56–59), which relied heavily on biblical geography, Islamic maps mentioned Qur'anic accounts of earthly creation only obliquely, and were more concerned with questions such as provincial administration and trade or pilgrimage routes.

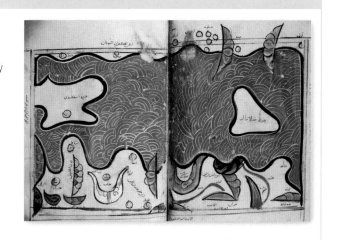

▶ **Each map in al-Idrīsī's atlas** depicted one latitude divided by one of 10 longitudinal divisions. This map shows the Indian Ocean.

Sawley Map

1200 ▪ MANUSCRIPT ▪ 29.5CM × 20.5CM (11½IN × 8IN) ▪ CORPUS CHRISTI COLLEGE, CAMBRIDGE, UK

UNKNOWN

SCALE

A small drawing used to illustrate *The Picture of the World*, a popular historical chronicle by theologian Honorius Augustodunensis (1080–1154), the *Sawley Map* is the earliest known English mappa mundi, or world map. It is based on biblical views of the size, shape, and origin of the world, and predates many more famous examples of this type of map. Beautifully illustrated with green seas, violet rivers, and red ridges representing surface relief, its 229 inscriptions and illustrations reflect a mixture of classical and Christian sources. It is oriented with east (oriens) at the top and west (occidens) at the bottom. Unlike many later mappae mundi, the map's centre is not Jerusalem but the island of Delos, the heart of ancient Greek mythology. Europe is relatively unimportant.

Despite its classical influences, the map's main emphasis is biblical history. Moving from the Garden of Eden at the top of the map, down to the Tower of Babel in the upper centre, through the Holy Land, including the boundaries between the Twelve Tribes of Israel and various locations in the life of Jesus, this is a map of religious affirmation. Guarded by four hovering angels, the *Sawley Map* depicts a Christian world awaiting the Day of Judgment.

IN CONTEXT

Records show that the *Sawley Map* was owned by – and possibly made in – Sawley Abbey, a monastery in England. Many early Christian mappae mundi were made in monasteries, which had the wealth and scholarly manpower to produce such work. These maps presented a powerful image of the world according to Christianity, which the Church disseminated. As well as showing places, their imagery depicted a series of ages, from the Old Testament to the medieval world.

▶ **The Psalter World Map** (1265), an illuminated miniature, shows Christ blessing the world, which has Jerusalem at its centre.

Visual tour

KEY

▶ **ANGEL AND GOG AND MAGOG**
The angel in the top left points towards Gog and Magog, monstrous mythical figures that are unleashed during the conflict preceding the Day of Judgment. The angels are similar to those in the Book of Revelation in the New Testament, who hold back destructive winds, suggesting that this is an image of Christian righteousness triumphing over the world's sinful chaos and destruction.

`1`

`2`

▲ **EGYPT AND ETHIOPIA**
Information at the map's margins becomes increasingly vague, resorting to garbled classical beliefs and Christian miscellany: Africa shows burning mountains and reptilian basilisks alongside St Anthony's Monastery and Egypt's Pyramids, named "Joseph's Barns" after the Old Testament figure who it was believed stored grain in them.

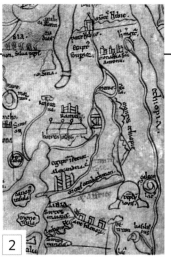

`3`

▲ **DELOS AND SCYLLA AND CHARYBDIS**
The depiction of the Mediterranean is indebted to Greek mythology. As described in Homer's *Odyssey*, the monster Scylla and the whirlpool Charybdis guard the straits between Italy and Sicily (not part of Italy at that time). The island of Delos, centre of the map and said to be the birthplace of Greek gods Apollo and Artemis, is visible to the left.

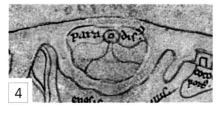

`4`

▲ **PARADISE** The most important location on the map is the Garden of Eden. It is shown exactly as described in Genesis: "A river flowed out of Eden to water the garden, and there it divided and became four rivers".

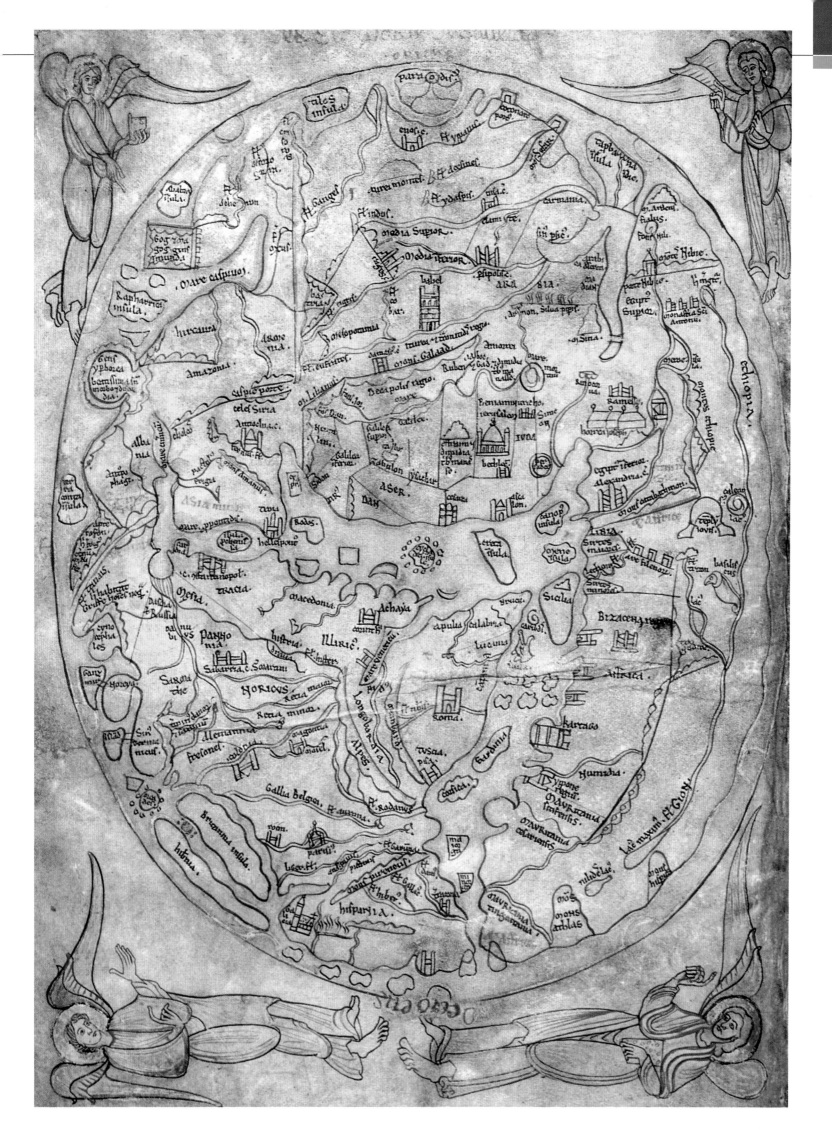

Carte Pisane

c.1275–1300 ▪ PARCHMENT ▪ 50CM × 1.04M (1FT 8IN × 3FT 5IN) ▪ BIBLIOTHÈQUE NATIONALE, PARIS, FRANCE

UNKNOWN

SCALE

The *Carte Pisane* is the earliest known portolan chart, making it one of the most important maps in the history of mapmaking. Portolan charts (the name comes from the Italian *portolano*, meaning "port" or "harbour") are simple maps of navigational routes across water, drawn using compass directions (see pp.68–71). Although it is the oldest of its type to survive, the *Carte Pisane* is almost identical in detail and execution to much later examples, almost as though the technique of producing this kind of chart emerged out of nowhere. The chart's author, original owner, and precise use remain a mystery, and its name simply refers to its much later discovery in Pisa, Italy. Even its exact age is contested, although most experts believe that it dates from around the late 13th century.

A tool for navigation

The *Carte Pisane*'s poor condition indicates that it was well used by navigators sailing across the Mediterranean, which is its main focus. It also depicts a sketchy British Isles at top left, the Holy Land at far right (a cross at Acre is the only obvious religious symbol), the Black Sea, and the Atlantic coast. It contains some 1,000 place names in a variety of languages, and the mapmaker tried to impose scientific order in pursuit of navigational accuracy by inscribing these names with care, at right-angles to the coast, and by including a network of compass directions and geometrical grids.

> [It] arrives like a **bolt out of the blue, revealing a new kind of world,** fully formed, and entirely **without precedent**

TOBY LESTER, *THE FOURTH PART OF THE WORLD*

Visual tour

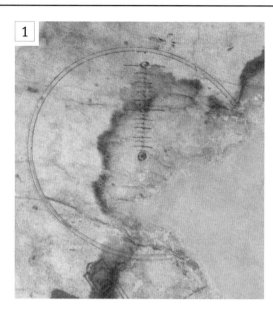

KEY

▶ **SCALE BAR**
Remarkably for its time, the map contains a scale bar, shown within a circle drawn using a compass and dividers. The scale is divided into segments of nautical miles, although at this time mariners rarely used scale. Despite the innovation here, later portolan charts abandoned the scale bar.

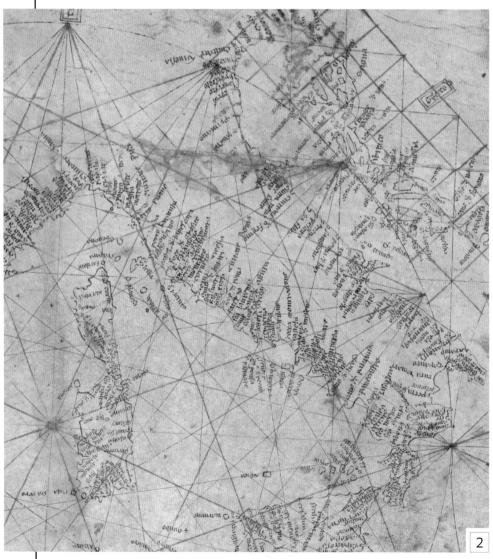

▲ **ITALIAN COASTLINE** Although the map was probably made in Italy and is relatively accurate in its depiction of the Mediterranean coastline, the Italian peninsula is poorly drawn, in particular the shape of the "heel". However, the map does accurately name many Italian ports and cities.

3

▲ **GRID OVER NORTH AFRICA** The cartographer appears to have used a compass and ruler to create so-called "four-by-four" grids, which divide a distance of about 320km (200 miles) into four equal parts. This represents an extremely ambitious attempt to use mathematics to impose a linear scale on the map to assist in navigation.

◀ **SARDINIA** The chart shows Sardinia and other Mediterranean islands in great detail, because sailors on commercial sea routes used them to orient themselves. The island's outline is almost as good as it would be on a modern map, and contains all the relevant ports and towns.

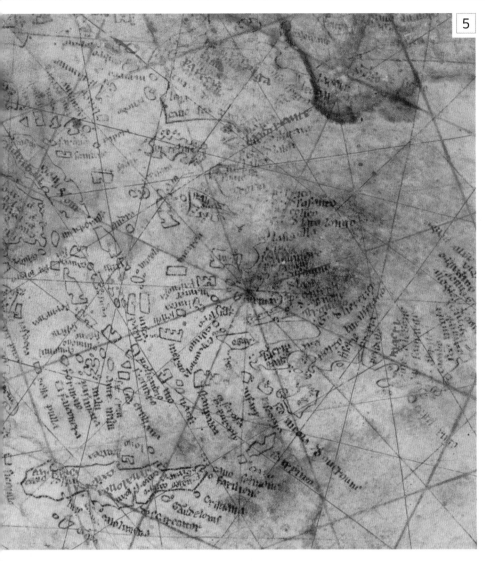

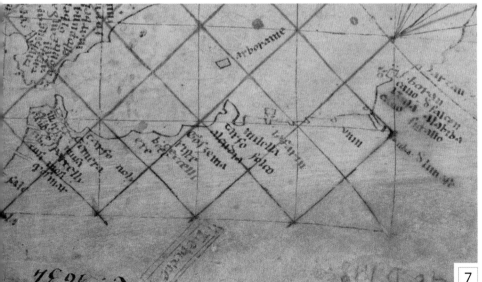

5 ◄ **COASTLINE OF ASIA MINOR** One of the map's most crowded areas shows Asia Minor and the Black Sea. This was the region controlled by the Byzantine Empire, although there is little sign of its presence here. Portolans were concerned only with water and coastlines: the interior and human geography held little or no interest for the mapmaker.

▼ **BRITISH ISLES** The mapmaker drew Britain as a crude, irregular shape, and set it on an inaccurate east-west axis, revealing his lack of interest in the island's commercial importance. The map names only six British places, including London ("Londra"), lying on what looks like the River Thames at the bottom centre.

6

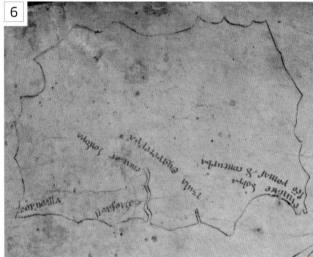

ON **TECHNIQUE**

Portolan charts were drawn using circles, lines, and grids based on repeated readings of compass directions. Probably the first things drawn on this chart were two large circles, one at each end of the Mediterranean, each divided into 16 wind directions based on identifying north using a compass. The mapmaker then drew the coastlines before adding place names, which were located by rechecking the compass directions.

▲ **This 1593 portolan chart created by Joan Oliva** has stylistic similarities to the much earlier *Carte Pisane*.

7 ▲ **NORTH WEST AFRICAN COASTLINE** The grid imposed on northwest Africa also contains one of the first sketches of the Moroccan coastline, including Azzemour, just below modern-day Casablanca. Its inclusion anticipates by over a century the Portuguese voyages down this stretch of coastline. Unlike contemporary medieval mappae mundi, there are no monsters in this portolan's Africa.

Hereford Mappa Mundi

c.1300 ▪ VELLUM ▪ 1.58M × 1.33M (5FT 4½IN × 4FT 4¼IN) ▪ HEREFORD CATHEDRAL, HEREFORD, UK

SCALE

RICHARD OF HALDINGHAM

With nearly 1,100 inscriptions detailing the geographical, theological, cosmological, and zoological dimensions of medieval life, the *Hereford Mappa Mundi* is one of the period's greatest surviving artefacts, offering a unique glimpse into the European Middle Ages. The map was designed in around 1300 by a team of clergymen based at the cathedrals of Lincoln and Hereford, England, led by the enigmatic Richard of Haldingham. It was made from vellum (calfskin), with the neck at the top and the spine running down the middle. The skin was first cured and scraped before a team of scribes and artists began work decorating its inner side using ink, gilt, and various pigments. Since its creation, it has remained in Hereford Cathedral, England, where it can still be seen today. The map's meaning and function have puzzled scholars for centuries: some regard it as part of the Christian Church's role as a source of knowledge and guidance, educating the congregation in the geographical and theological mysteries of the world; while others believe it contains more mundane references to local disputes in the medieval Hereford diocese.

A map to Paradise

This map shows the world according to the beliefs of medieval Christianity. It is oriented with east at the top, and is decorated with scenes from ancient and classical history, and stories taken from both the Old and New Testaments. These are shown as the eye moves westwards down the map, ending in the western Mediterranean. Jerusalem lies at the exact centre of the map, with Christ's crucifixion shown above it, while grotesque animals and monstrous peoples appear at its margins, products of the medieval Christian imagination. The figure of Christ stands above it all, watching the drama of the Day of Judgment. This is a map that leads the reader to a spiritual destination – Christian heaven – rather than an Earthly one.

RICHARD OF **HALDINGHAM**

c.1245–c.1326

The lower left-hand corner of the map contains an inscription claiming that it was created by "Richard of Haldingham and Lafford" – who, it says, "made and planned" it.

"Richard of Haldingham and Lafford" is a shadowy figure, also referred to elsewhere as "Richard de Bello", about whom comparatively little is known. Originally hailing from the town of Battle in Sussex, he began his career as a clergyman in the adjoining Lincolnshire parishes of Haldingham and Lafford (known today as Holdingham and Sleaford), and rose to become first canon, then treasurer, at Lincoln Cathedral. Around the time of the map's creation, Richard of Haldingham was promoted to more powerful religious positions in the Herefordshire area. In this capacity, he may have become leader of a team of clergymen who designed the map for the glory of Hereford and its cathedral.

Let all who have this history, Or shall hear or read or see it, Pray to Jesus in His Divinity, To have pity on Richard of Haldingham and Lafford, Who has made and planned it

Visual tour

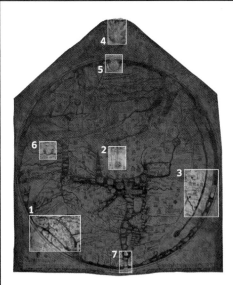

KEY

▶ **BRITISH ISLES** Squashed into the bottom corner of the map, the British Isles are divided between "Anglia", "Wallia", "Scocia", and "Hibernia" – England, Wales, Scotland, and Ireland respectively. Most of the 81 inscriptions name rivers, cathedral cities, and tribes. London, Edinburgh, and Oxford are mentioned, as are places central to the map's creation. Lincoln and its cathedral are shown, as are Caernarvon and Conway, both sites of English garrison towns built in the late 13th century, in the Welsh Marches near Hereford.

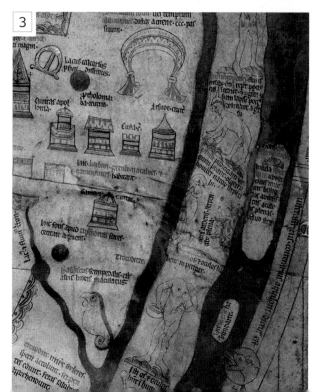

1

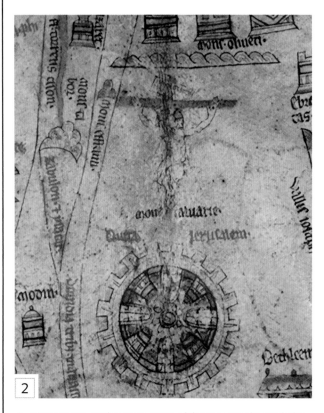

2

▲ **JERUSALEM** At the exact centre of the map sits the symbolic centre of Christianity, the site of Christ's crucifixion – which is shown just above the city itself. Labelled in red, Jerusalem is shown as a circle ringed by 16 crenellations, containing eight towers, with the Church of the Holy Sepulchre at its heart. A hole at the centre of the circle, perhaps the first mark made by the mapmaker, was created with a compass.

▶ **MONSTROUS PEOPLES**
Mythical places, fabulous creatures, and monsters fill sub-Saharan Africa. From the top the map shows cave-dwelling "Troglodites", a poisonous "Basilisk", a race called "Blemmyes" with "mouths and eyes in their chest", and the "Philli" who "expose their newborns to serpents".

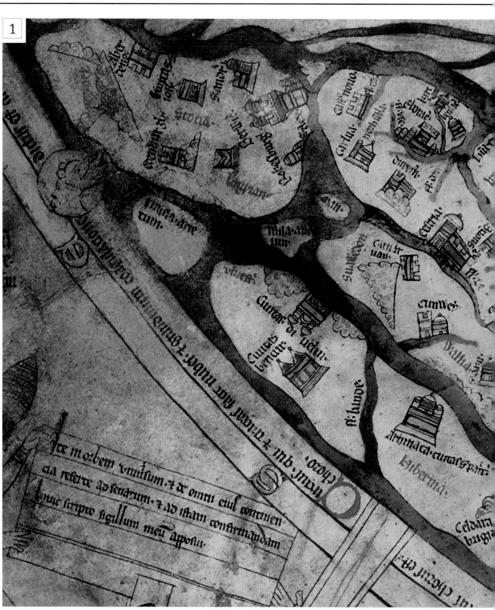

3

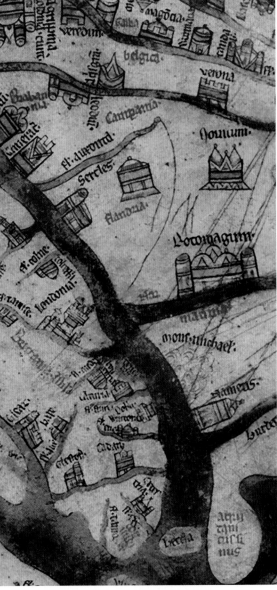

5

▲ **GARDEN OF EDEN** The northernmost point on the map shows one of the foundational locations in Christian belief, the Garden of Eden. Labelled as "Paradise", it shows Adam and Eve in the moment of original sin. The garden, set within a fortified wall and gates, is shown as the origin of four great rivers of the ancient world: the Phison, Gehon, Tigris, and Euphrates. Christian belief dictates that from this scene all subsequent human history flows.

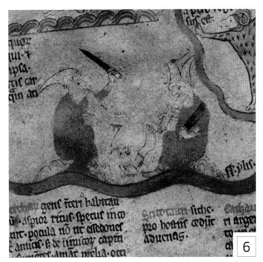

6

4

◀ **CHRIST** At the top of the map, outside terrestrial time and space, stands Christ. He displays the scars of the crucifixion, and presides over the Day of Judgment, announcing "Behold my witness", while Mary looks on in adoration. Behind him, the wavy lines of the heavens contrast with the earthly world being judged.

IN **CONTEXT**

Medieval Christian mappae mundi, including the Beatus map (below), reshaped geography along biblical lines, mixing a literal and allegorical interpretation of the Old and New Testaments to create a world map. They showed the passage of religious time, as well as places such as Paradise and figures including Adam and Eve. These maps nearly always anticipated the biblical Day of Judgment, and the world's end.

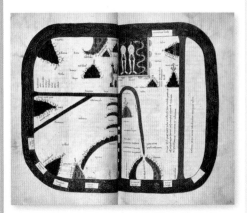

▲ **The 12th-century Beatus world map** shows Adam, Eve, and the serpent in the Garden of Eden.

◀ **CANNIBALISM** Since classical times, Scythia, the central Asian region north of the Black Sea, had been depicted as located at the end of the world, inhabited by monstrous, cannibalistic barbarians. Here the map shows a sinister, knife-wielding race called the "Essedones" who, once their parents die, eat them in a "solemn feast of animal meat mixed with human flesh". They are shown snacking on human heads and feet.

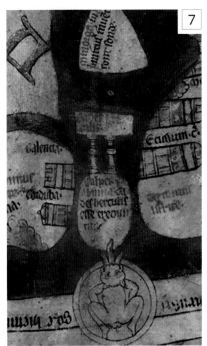

7

◀ **PILLARS OF HERCULES** The map's westernmost point shows the "Columns of Hercules" on the "Rock of Gibraltar". In classical mythology, one of Hercules's labours involved travelling here, the boundary of the known Greek world, where the Mediterranean gave way to the "dark sea" of the Atlantic. The pillars represent the limits of classical and Christian geography for the next century, until the Portuguese began to sail out into the Atlantic. The map also names "Gades" (modern-day Cádiz), west of Gibraltar.

DISCOVERY AND TRAVEL

- Catalan Atlas

- Kangnido Map

- Portolan Chart

- Frau Mauro's World Map

- Juan de la Cosa's World Chart

- Map of Venice

- Map of Imola

- First Map of America

- Piri Re'is Map

- Map of Utopia

- Map of Augsburg

- Universal Chart

- Aztec Map of Tenochtitlan

- New France

- A New and Enlarged Description of the Earth

1300–1570

Catalan Atlas

1375 ▪ VELLUM ▪ 65CM×50CM (2FT 1½IN × 1FT 7½IN) PER PANEL ▪ BIBLIOTHÈQUE NATIONALE, PARIS, FRANCE

SCALE

ABRAHAM CRESQUES

This is the finest known example of the medieval Catalan tradition of mapmaking, which centred on Barcelona and the Balearic islands. Created by Jewish cartographer Abraham Cresques and his son Jehuda, it was believed to have been made for the future French King Charles VI, and, significantly, it unified the styles of Christian mappae mundi and traditional Mediterranean portolan navigation charts for the first time. Spectacularly illuminated on four vellum panels that fold out like a screen, it utilized the skills of a specialist binder and illustrator, as well as a cartographer. Several distinctive features are immediately noticeable. Firstly, unlike most medieval mappae mundi, this one has no clear directional orientation – the

northern sections appear upside down when viewed from the south, suggesting it was supposed to be placed on a table and examined by walking around it. Then there is the sheer amount of detail and drama, including 2,300 names and many more mountains, rivers, cities, and animals. Finally, the map also lays claim to an important world first – the depiction of a compass rose.

Contemporary influences

The *Catalan Atlas* draws on three different areas of geographical knowledge: mappae mundi, from which it gets its sacred centre, Jerusalem, along with a wealth of other classical and biblical details; portolan charts, which

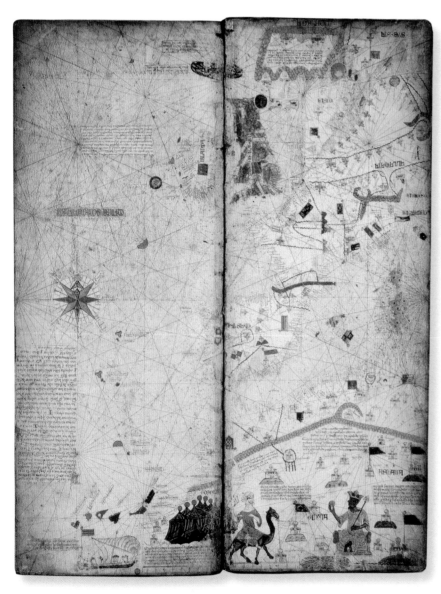

> A complete view of the then-known world, stretching from the newly discovered Atlantic islands to the China Sea. It is an indispensable summary of late medieval Europe's geographical knowledge, one of the last great mappae mundi

JEAN MICHEL MASSING, FRENCH ART HISTORIAN

provide the general shape and orientation of the two panels showing Europe and the Mediterranean, as well as the rhumb lines that crisscross the whole map; and the latest travel reports from Asia, in particular those of Marco Polo (1254–1324). In fact, Marco Polo's writing provided a completely new understanding of the Far East, which transformed the right-hand panels, although they are still riddled with tall tales of fabulous kingdoms and monstrous peoples. Overall, the map captures classical and Christian geographical beliefs retreating in the face of trade and exchange, shown stretching from the Saharan gold trade in the West to the pursuit of spices in the Far East.

ABRAHAM **CRESQUES**

1325-1387

Also known as "Cresques of Abraham", Abraham Cresques was a celebrated cartographer who worked for the King of Aragon, Pedro IV (1336-87). As well as making maps, Cresques also built clocks, compasses, and nautical instruments.

Born into a wealthy Jewish family living in Palma de Mallorca, Abraham Cresques often worked closely with his son, Jehuda. Both were part of the celebrated 14th-century Mallorcan school of mapmakers. In documents associating Cresques with the *Catalan Atlas*, he is referred to as "Cresques the Jew" and "a master of maps of the world and of compasses". Although he designed later world maps, they have all been lost. After Cresques' death, his son continued to make maps, but he was forced to convert to Christianity in 1391, and became known as Jaime Riba.

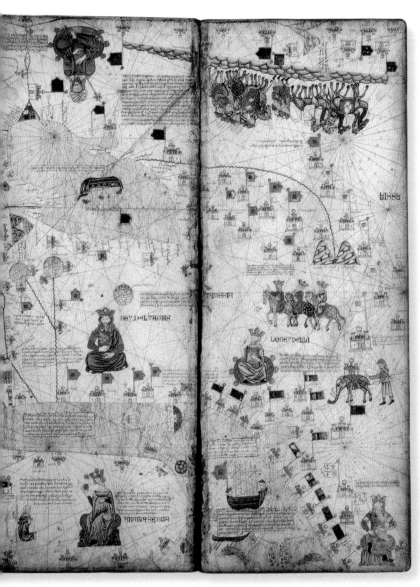

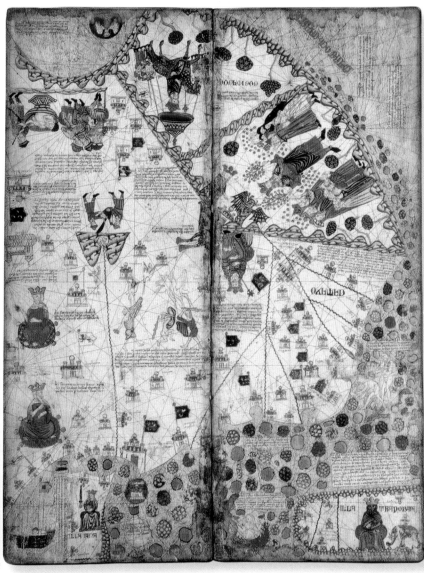

Visual tour

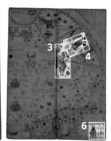

KEY

▲ **CANARY ISLANDS** Cresques renders the Canary Islands, situated between Spain and Africa, with remarkable accuracy, and only La Palma is missing. According to Cresques' inscription, the boat south of the islands belongs to his fellow Mallorcan, explorer "Jacme" (Jaume) Ferrer, who set out for the fabled West African "River of Gold" in 1346, but was never seen again.

1

▷ **CHRIST-LIKE KING** Maps produced for medieval Christian rulers often inaccurately portrayed Christian rulers in Africa, India, and Asia in the hope of halting the spread of other religions, such as Islam and Hinduism. Here, in the Far East of Cresques' atlas, an opulent Christian ruler holds court.

▽ **MARCO POLO'S CARAVAN** The map is one of the earliest to show unequivocally the influence of Marco Polo's *Description of the World*, an account of his epic travels through Asia between 1276 and 1291. This scene portrays a caravan including Marco and his family travelling along the Silk Route, accompanied by Mongol envoys.

2

▲ **AFRICAN WEALTH**
The lucrative trans-Saharan trade routes are represented by a Touareg nomad on a camel and Mansa Musa, "lord of the Negroes of Guinea". Mansa Musa ruled Mali from 1312 to 1337, and Cresques described him as "the richest and noblest king in the world", due to the region's gold resources.

▷ **ALEXANDER THE GREAT, DEMON, AND KUBLAI KHAN**
At the edges of the map, religion, classical beliefs, and travellers' tales collide. Here, ancient Macedonian King Alexander the Great gestures towards a devil, conflating the biblical figures of Gog, Magog, and the Antichrist. Below in green, viewed from the other side of the map, is Kublai Khan, founder of the Yuan dynasty.

3

5

6

7

▲ **"TAPROBANA"** Taprobana was an island that caused classical and medieval writers terrible confusion. Some believed it to be what is now Sri Lanka, while others thought it was Sumatra. The latter view was favoured by Marco Polo and Cresques, who put it in southeast Asia, in a sea of 2,700 islands, ruled by an exotic, elephant-loving king.

▲ **THE THREE MAGI** Christian tradition suggests the Three Magi that visited Jesus after his birth came from Arabia, Persia, and India. Here they appear looking suspiciously European, riding across northern India, en route to Bethlehem. It is another of the map's typically anachronistic fusions of past and present.

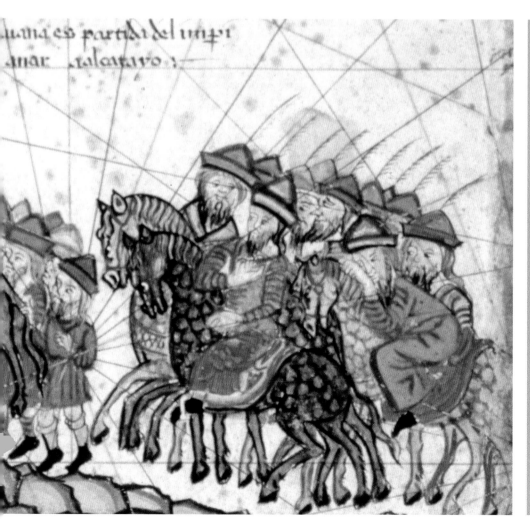

IN **CONTEXT**

The *Catalan Atlas* calls itself an "image of the world", and it presented more than just geography. Two additional panels also offer a cosmology of the known world, featuring astronomical and astrological text and diagrams. One panel, below, has a cosmographical picture showing concentric planetary circles with Earth at its centre, a geocentric belief inherited from the Greeks and given a Christian interpretation here and throughout the atlas.

▲ **This large, circular chart** is framed by images of the four seasons and has information about the zodiac, the seven known planets, and constellations.

Kangnido Map

c.1402 ■ INK ON PAPER ■ 1.58M × 1.63M (5FT 2¼IN × 5FT 4¼IN) ■ RYUKOKU UNIVERSITY, KYOTO, JAPAN

KWŎN KŬN

SCALE

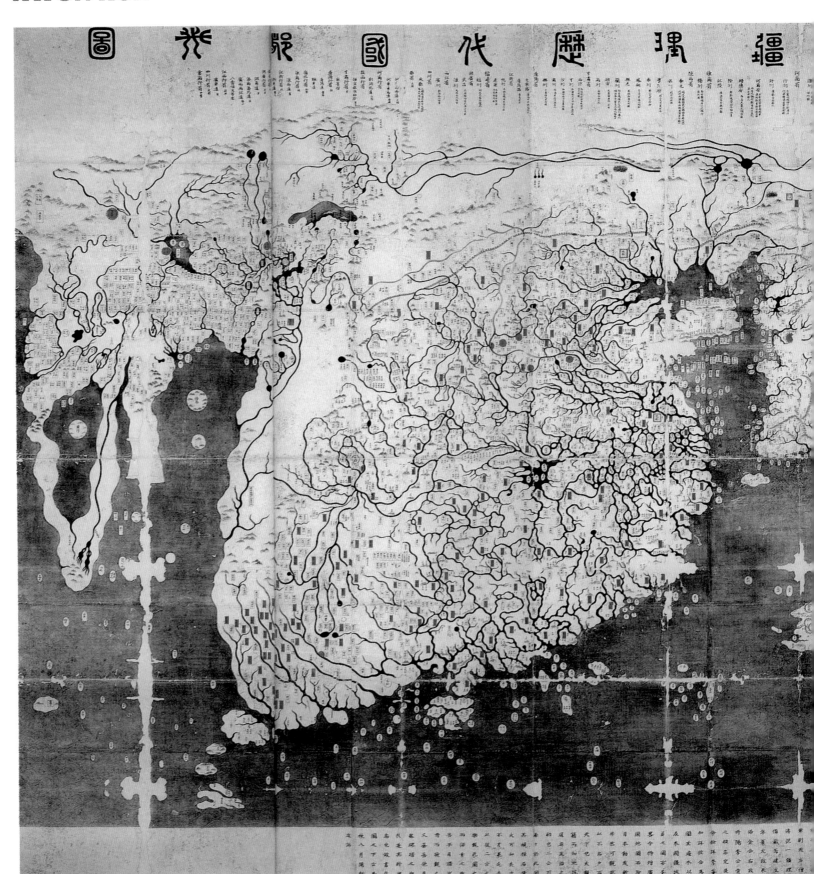

Korea's most famous map, the *Kangnido* (a Korean abbreviation of its full title, translated as "Map of Integrated Lands and Regions and of Historical Countries and Capitals"), was first made in 1402. It was then lost, although several copies, including this one from around 1560, were made later. It is the earliest known East Asian world map, and the first to depict the new Korean Goryeo Empire as well as Europe. As the map's creator, Kwŏn Kŭn, explains in the text at the bottom, this is a world dominated by the Chinese Ming dynasty. It also depicts a vastly inflated Korea at the top right, shown as being three times the size of Japan (bottom right), which in reality is larger. The rest of the world is peripheral: India almost completely disappears, although Europe and

Africa are shown in surprising detail considering there is no evidence for the spread of Greco-Roman geographical knowledge as far as Korea. At the top of the map the landmass stretches to infinity, suggesting that, like many Chinese mapmakers, Kwŏn Kŭn believed that the heavens were round but the Earth was square and flat.

KWŎN **KŬN**

c.1352–1409

A Neo-Confucian scholar, diplomat, and poet, Kwŏn Kŭn was also a key administrative figure in both the Goryeo and Joseon Korean dynasties in the 14th and 15th century.

Kwŏn Kŭn lived during a time of great dynastic upheaval, and he spent some time in exile for his support of one of the competing factions. He travelled on sensitive diplomatic visits to Ming-dynasty China, wrote various books on education and ritual, and spent the last decade or so of his life working on his famous world map.

Visual tour

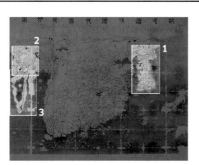

KEY

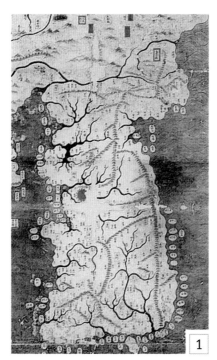

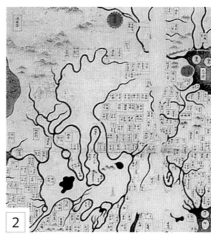

2

▲ **EUROPE** The Mediterranean is shown coloured in white. The Italian peninsula is in the middle, above an island, possibly Sicily, with the red dot near the top probably showing Constantinople (Istanbul), the capital of the Byzantine Empire. Alexandria and even Germany are shown, exhibiting the extent of Korean geographical knowledge.

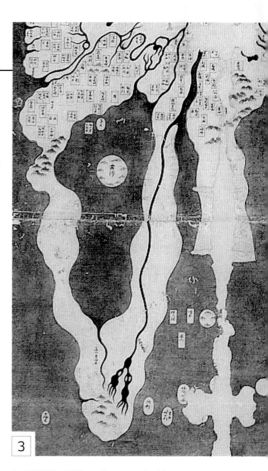

3

▲ **AFRICA** With a circumnavigable southern cape and a central section believed to represent the Sahara desert, Africa is shown here with greater accuracy than on contemporary European maps. This suggests that navigators from Asia had actually managed to sail around the continent.

◄ **KOREA** With its distinctive vase-shaped peninsula, Korea is far too large here, even in comparison with China. The most important features are its arterial network of rivers and mountains, as well as naval bases, which are shown dotted along the coast, resembling islands.

1

Portolan Chart

1424 ▪ INK ON VELLUM ▪ 57CM × 89CM (1FT 10½IN × 2FT 11IN) ▪ JAMES FORD BELL LIBRARY, MINNESOTA, USA

SCALE

ZUANE PIZZIGANO

The most mysterious map in the history of Western exploration, Zuane Pizzigano's modest portolan chart of the north Atlantic Ocean is an enigma. It was discovered in 1853, but we know nothing about where it came from; its maker is a mystery, and even his signature is questionable. A typical portolan (see p.53), the chart is drawn with coloured ink on vellum, using rhumb lines and a scale bar to provide navigational assistance for people sailing along the coasts of Ireland, England, France, the Iberian Peninsula, the Balearics, and islands off northwest Africa.

The chart depicts the Atlantic Ocean, but excludes most of the Mediterranean Sea. It names 543 places, nearly all along the coastline, including a number of mysterious islands in the Atlantic – most significantly the legendary Antilia island, which it names for the first time. The island, which appears on many subsequent maps, does not correspond to any known archipelago. By 1419, the Portuguese had discovered only Madeira, yet this chart seems to show far more – possibly even America, almost 70 years before Columbus.

> One of the most **precious** monuments in the history of cartography and of geography

ARMANDO CORTESÃO, PORTUGUESE MAP HISTORIAN

ZUANE **PIZZIGANO**

C. EARLY 15TH CENTURY

The chart's author is almost as mysterious as its content. There is no record of Zuane Pizzigano's birth or death, and the portolan chart is the only map made by him to survive.

He is believed to have been a Venetian, an assumption based on the Italian dialect used in inscriptions across the chart. Pizzigano's unusual first name also lends support to the claim that he was Venetian – Zuane is a variant of Giovanni that was sometimes used in Venice but is very uncommon elsewhere. He may also have been a descendant of a 14th-century family of cartographers called the Pizigani, although there is no evidence to confirm this.

The legendary
island of Antilia

The island
of Himadoro

Visual tour

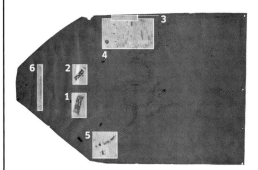

KEY

▶ **PHANTOM ISLAND OF ANTILIA** The huge, rectangular island labelled "Antilia" appears in the Atlantic for the first time on this chart. Although it features seven completely fictional cities, some believe this to be a Caribbean island, or even part of the American coast.

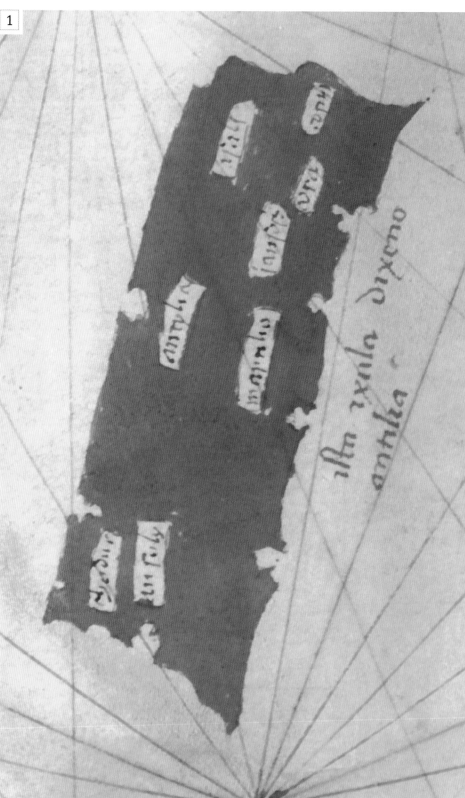

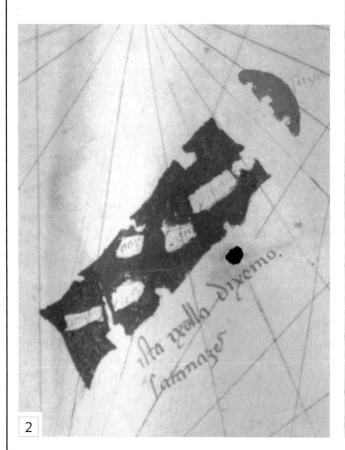

▲ **THE ISLE OF DEVILS**
North of Antilia, below the umbrella-shaped Saya, is another unknown island, labelled "Satanazes" and "Isle of Devils". Some historians speculate that its legendary devils derive from Norse myths about Greenland, and that Satanazes is Labrador.

▶ **SCALE BAR** The chart's rudimentary scale bar, at the top, is divided into eight parts further subdivided into fifths. Each of these subdivisions corresponds to 10 miles (16km). Like many other 15th-century charts of the Atlantic, the portolan's scale is approximately 1:6,500,000.

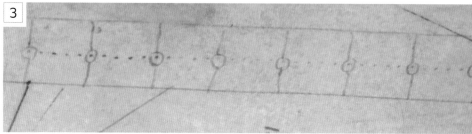

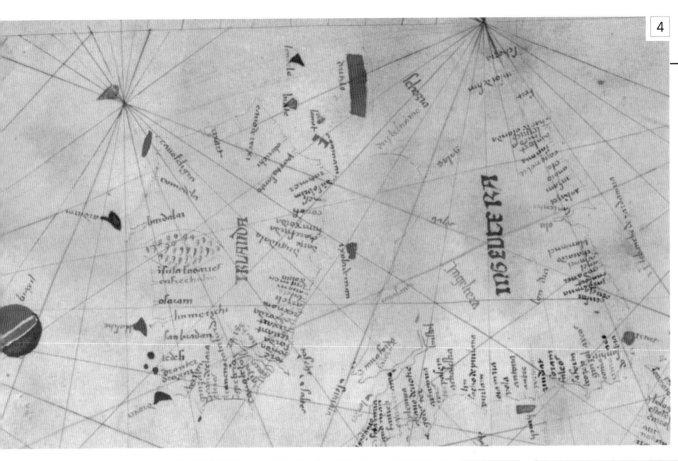

4

◄ **ENGLAND, IRELAND – AND "BRASIL"?** West of England ("Ingeutra") and Ireland ("Irlanda") is a circular red-blue island marked "Brasil". This was a mythical island found on European maps from the early 14th century and bears no relation to the South American Brazil (discovered in 1500). The legendary Brasil probably derived from the island of eternal happiness, Hy Breasal, in Gaelic Irish folklore.

IN **CONTEXT**

Like many 15th-century mapmakers, Pizzigano was unsure about the size and extent of the globe and had to assess the truth of contradictory travellers' tales to map islands described in equal measures of fact and legend. This was a time, prior to European discovery of the Americas, when the Atlantic was regarded as a dark and shadowy ocean from which sailors rarely returned. Some Christian mapmakers showed distant lands populated by monsters, dragons, and entrances to heaven or hell. Armchair geographers such as Pizzigano tried to locate places that were thousands of miles away, enabling fanciful enthusiasts to speculate that such maps showed the Caribbean, America, China, and even Atlantis.

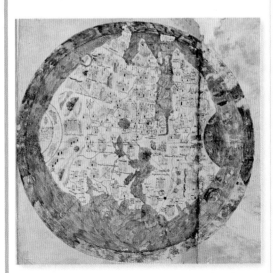

▲ **This world map is from** the *Atlante Nautico* of 1436, produced by the Italian sailor and mapmaker Andrea Bianco.

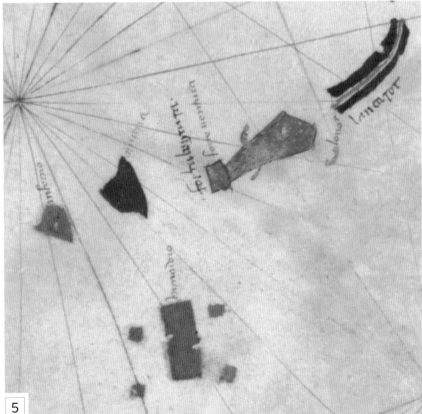

5

▲ **HIMADORO: A REAL DISCOVERY?**
To the south of the Canaries is the island labelled "Himadoro", surrounded by four smaller islets. The name may be a garbled version of "The Island of Gold", and the islands a very early reference to the Cape Verde archipelago, even though the Portuguese only officially announced its discovery by Diogo Gomes in 1460.

▶ **LAYING CLAIM TO THE MAP**
In the neck of the map on the far left is an inscription which reads, "On 22nd August 1424, Zuane Pizzigano made this map". Even this attribution is unclear, as the surname is smudged – perhaps deliberately, in an attempt to erase it.

6

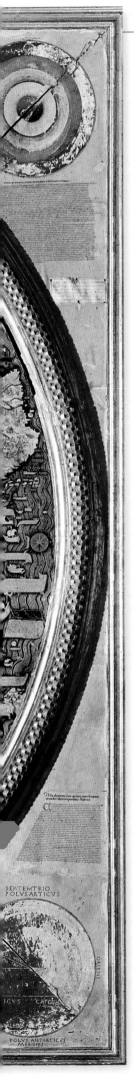

Fra Mauro's World Map

C.1450 ■ VELLUM ■ C.2.4M × 2.4M (C.7FT 11IN × 7FT 11IN) ■ MUSEO CORRER, VENICE, ITALY

SCALE

FRA MAURO

Fra Mauro's monumental mappa mundi is one of the most beautiful and important works in the history of cartography. Created on the Venetian island of Murano by the monk Fra Mauro, it represents a complete picture of the cosmos according to late medieval Christian belief. It also marks the beginning of the end of early medieval mappae mundi that reflected biblical geographical teaching. It cautiously embraces new scientific techniques and discoveries, even where they question the assumptions of established classical and Christian geography.

Faith and science

Although the world is represented here in circular form, it should be understood as part of the wider cosmological scene encased within the square frame, showing Paradise, the known universe, and astronomical diagrams. One of the map's most striking characteristics is that it shows south is at the top, rejecting both Ptolemy's northern orientation (see pp.24–27), and the eastern orientation of virtually all other medieval mappae mundi (see pp.56–59). In a break from convention, Jerusalem is not shown as the centre of the world, and the coastline of southern Africa is also shown as circumnavigable, a further rejection of classical beliefs (although intriguingly, this is shown more than 30 years before the Portuguese first rounded the Cape of Good Hope in 1488). The map's 3,000 inscriptions incorporate the latest

traveller's reports from Africa and Asia to produce some of the era's most detailed accounts of both continents, including the first recorded depiction of Japan by a Western cartographer.

Created on a huge scale using the finest vellum, gold, and pigments, a team of Venice's greatest cartographers, artists, and copyists worked alongside Fra Mauro to make an object that was not only expensive (costing an average copyist's annual salary), but also a beautiful work of art that captures geographical knowledge from the perspective of 15th-century Venice.

FRA **MAURO**

c.1400–c.1464

Fra Mauro was a lay member of the Camaldolese monastery of San Michele di Murano, Venice.

He appears in the monastery's records in 1409, and remained employed there until 1464, the probable date of his death. His main job was recorded as collecting the monastery's rents, but from the 1450s he is also named as the maker of a series of mappae mundi. The first, for "the benefit of the Republic of Venice", was made for the Portuguese court, who took a keen interest in what Fra Mauro knew of the Far East; he also made a later copy of the same map, which does not survive. His fame earned him the title of "cosmographer without equal".

I have been… **investigating** for **many years** and frequenting persons **worthy of faith** who have seen **with their own eyes** what I **faithfully** report ""

FRA MAURO

Visual tour

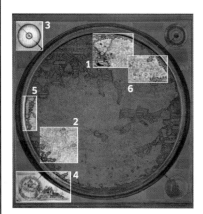

KEY

▶ **CHINA** In contrast to the popular Western view of the time, this map does not feature exotic images of fabled rulers and the people of the East, and instead depicts a dense network of Chinese cities and trading ports. These were based on a careful fusion of classical sources and more recent traveller's reports, including Marco Polo's – from which Fra Mauro gets the names of 30 cities – and the more recent accounts of Niccoló de' Conti (1395-1469). The Yangtze and Yellow rivers are shown, as are key places such as Manzi, Quinsai, and Zaiton. Fra Mauro's sources were mainly preoccupied with trade.

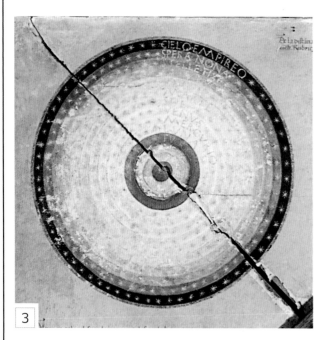

▲ **COSMOS** The celestial world is painted using dazzling gold and a blue pigment called azzuro. It is composed of 10 concentric spheres, with the Earth at its centre. This geocentric belief was established by the Greeks and developed by Christianity's "holy theologians", who are discussed on the map. The spheres move outwards from the Moon, Mercury, Venus, the Sun, Mars, Jupiter, and Saturn, before reaching the fixed stars, a crystalline sphere, and finally the "Empyrean", or firmament, of what Fra Mauro calls "one heaven", and the realm of God.

▶ **SOUTHERN AFRICA** Fra Mauro shows a circumnavigable Africa - something unknown to the ancients, including Ptolemy - more than 30 years before the Portuguese rounded the Cape of Good Hope in 1488. The annotation claims that "around 1420 a ship or junk from India" sailed round what Fra Mauro labels the "Cape of Diab". This suggests that he knew of the voyages of the Ming Dynasty's admiral Zheng He (see pp.134–37), who may have reached the Cape before any European.

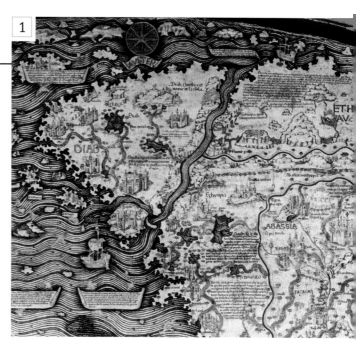

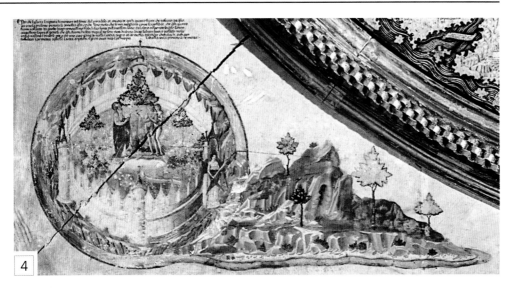

4

▲ **PARADISE** Christian mapmakers were obliged to put the biblical Paradise on a world map, even though they believed it existed outside mortal time and space. Fra Mauro's ground-breaking solution was to put Paradise outside the map's frame, adjacent to its putative location in the east. It shows God commanding the naked Adam and Eve, the origins of the Earth's four rivers, an angel anticipating the Fall, and the rocky, inhospitable, terrestrial world the couple are about to enter. It shows Paradise as separate from, but connected to, the inhabited Earth.

▼ **JAVA AND JAPAN** At the furthest limit of the map, right against the frame, Fra Mauro depicts the important spice-producing island of Java and, just below it, a much smaller island labelled "Ixola de Cimpagu". This is Japan, named on a European map for the very first time. The island shown is possibly Kyushu, the most southwesterly of Japan's four main islands.

5

IN **CONTEXT**

Venice was one of the great centres of medieval and Renaissance mapmaking. A great maritime power, it sponsored the creation of portolan charts (see p.53) used in navigation across the trade routes of the Adriatic and Mediterranean Seas. As a crossroads of trade, travel, and art between East and West, the city possessed the latest commercial and political news from locations as distant as China, which allowed it to stay at the forefront of international trade and diplomacy. This also gave its cartographers unprecedented access to information for making sea charts, regional maps, and mappae mundi.

▲ **In the 15th century, Venice was** well known as a centre for bustling maritime and commercial activity.

6

▲ **WEST AFRICA** Portuguese voyages down the African coast in the 15th century afforded Fra Mauro with detailed descriptions of human and physical geography. One annotation claims he used maps from local "clerics who, with their own hands, drew for me these provinces and cities". The African coast is no longer peripheral to the Christian imagination, and instead becomes a place of important trade relations en route to the east.

Juan de la Cosa's World Chart

1500 ▪ PARCHMENT ▪ 96CM × 1.83M (3FT 1¾IN × 6FT) ▪ MUSEO NAVAL, MADRID, SPAIN

SCALE

JUAN DE LA COSA

This lavishly illustrated world chart was the earliest map of the Americas to show Christopher Columbus's historic landfall in 1492. It was lost for centuries until it was discovered in its current fragile state in a Parisian shop in 1832. In 1853, it was bought by the Queen of Spain, as it was originally intended to show the 15th-century Spanish crown the newly discovered territories to the west.

Some of the descriptions appear upside down or at a 90-degree angle, and the chart is also split down the middle, which suggests that it was meant to be displayed on a table and examined from all sides. The eastern half reproduces the geography of Europe, the Mediterranean, Africa, and Asia, based on Ptolemy (see pp.24–27) and medieval mappae mundi. The western half, which is on a much larger scale, draws on newer discoveries made by the Portuguese and Spanish in the Atlantic.

Mapping a new continent

De la Cosa was the first pilot and cartographer to portray the Americas as a distinct landmass, rather than as islands connected to Asia, as Columbus believed. On the left of this chart the new continent curves, green and inviting, but nothing explicitly associates it with Asia. To the north,

de la Cosa shows Newfoundland, visited by the English explorer John Cabot in 1497, while to the south he draws an emerging South American coastline, including Brazil, which was claimed by Spain and Portugal in the spring of 1500, just months before this chart was made. The map is defined by the

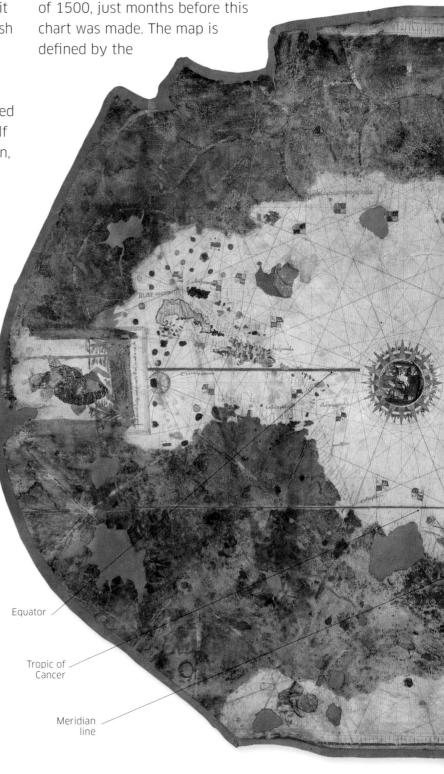

Equator

Tropic of Cancer

Meridian line

JUAN **DE LA COSA**

c.1450–1510

Born in Spain, Juan de la Cosa was a pilot and navigator on some of the earliest voyages to the "New World".

He sailed to the Americas on seven occasions, and was central to three of Columbus's famous voyages, including the first landing in the Bahamas in October 1492. On Columbus's third voyage, in 1494, de la Cosa seems to have questioned the assumption that Cuba was part of Asia. In subsequent voyages he sailed with the conquistador Alonso de Ojeda, who named Venezuela, and Amerigo Vespucci, who gave his name to the continent of America. De la Cosa was one of the first Europeans to set foot in South America, and he also explored Colombia, Panama, Jamaica, and Hispaniola. On his last voyage to Colombia, in 1509–10, he was killed by a poisoned arrow fired by local natives during a skirmish with Spanish troops.

One of the most important of all cartographic records of the early European exploration of the Americas

JAY LEVENSON, DIRECTOR, INTERNATIONAL PROGRAMME, MUSEUM OF MODERN ART, NEW YORK

horizontal *circulo equinocial* (equatorial line), the *circulo cancro* (Tropic of Cancer) to the north, and the green *liña meridional* (meridian line) running north to south. This line was politically demarcated at the Treaty of Tordesillas in

1494: everything to the west went to Spain; everything to the east to Portugal. On the right side of the chart, de la Cosa reverts to traditional geography, showing monsters and marvels; to the left, a new world begins to take shape.

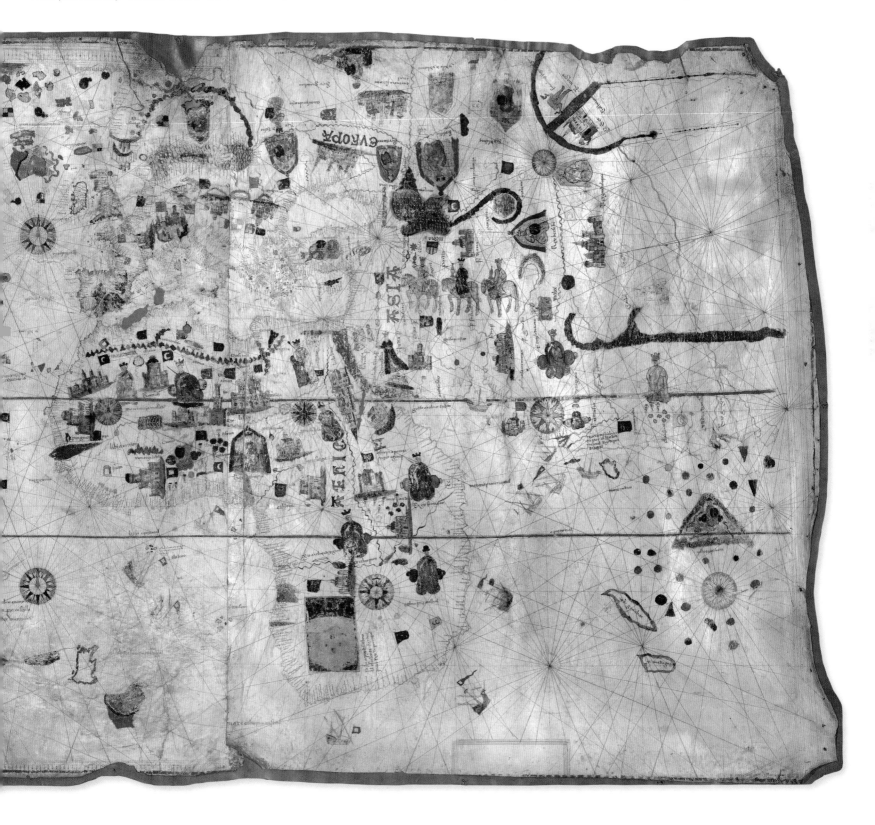

Visual tour

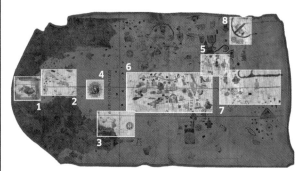

KEY

▶ **ST CHRISTOPHER** West of Cuba, de la Cosa has created an illustration of St Christopher carrying the infant Jesus across a river. The association with Columbus is obvious: like his namesake he travelled across the water, spreading the word of Christ to new lands and peoples. The inscription reads "Juan de la Cosa made it in the port of Santa Maria [near Cadiz] in the year 1500".

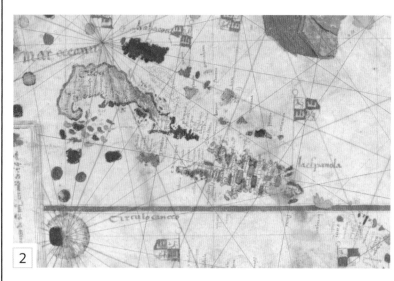

▲ **CARIBBEAN ISLANDS** De la Cosa has rendered Cuba, Hispaniola, and the Bahamas with great accuracy, probably thanks to the amount of time he spent sailing though the Carribean. Cuba is named on a map for the first time, shown in remarkable detail, and is clearly not part of Asia, as Columbus insisted. The island includes 27 locations, although like Hispaniola, it is shown incorrectly north of the Tropic of Cancer. Between the two flags northeast of Cuba is the island of "Guanahani", named "San Salvador" by Columbus and also believed to be where he first set foot in the Americas on 12 October 1492.

▶ **BRAZIL** In January 1500, the Spaniard Vicente Pinzon landed on the Brazilian coast, followed three months later by the Portuguese Pedro Cabral. De la Cosa credits Pinzon with discovering a "cape", and Cabral an "island". The 1494 meridional line gives Spain most of Brazil and South America, although here the shape of the land is wrong and Brazil is on the same latitude as Cape Town.

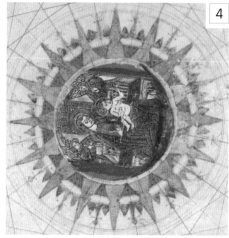

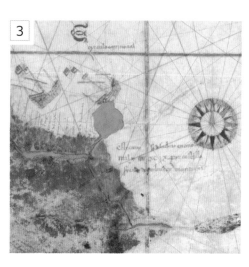

◀ **COMPASS ROSE** The 32-point compass rose, illustrated off the Brazilian coast, combines science with religion. Also known as a wind rose, it takes the four "cardinal winds", or directions, which are then further broken down into "intermediate winds", used to chart direction in navigation. At the centre of this example is an image of the Holy Trinity, demonstrating the religious imperatives behind the creation of this map.

5

◄ ASIA AND THE THREE MAGI
To the west, the map enthusiastically records new discoveries, but to the east it reiterates established classical and biblical beliefs. Next to "Asia", three oversized Magi head towards Syria bearing gifts, and just below them stands the Queen of Sheba with a drawn sword. Despite this religious symbolism, Jerusalem is not placed at the map's centre, unlike most mappae mundi (see pp.56–59).

◄ WEST AFRICAN COAST
The West African Gold Coast is the most accurately mapped area on the chart, probably as the region had been intensively explored and settled by the Portuguese since the 1420s. In typical portolan style, names are written at right angles to the coast, while the interior is decorated with various rulers and settlements, notably São Jorge da Mina, a Portuguese fort built in 1482.

6

IN **CONTEXT**

De la Cosa's chart marks an important shift from the use of portolans (see p.53) for seaborne navigation. Portolans were used for centuries in the Mediterranean, alongside compasses and knowledge of coastlines, islands, winds, and depths. They did not need to take into account the curvature of the Earth, because over relatively short distances such factors made little difference. However, from the late 15th century, as Portuguese and Spanish pilots sailed out into the Atlantic beyond sight of land, they needed a new type of chart to assess latitude, longitude, and distances covered, and that incorporated the globe as a sphere. De la Cosa's map reflects this period of transition. Although it lacks a graticule (a grid of latitude and longitude), it introduces meridional and zonal lines, as well as compass roses with radiating lines of orientation and direction, all of which were crucial for successful oceanic navigation.

▲ **Christopher Columbus's** discoveries heralded a new age in maritime navigation.

▼ INDIA The depiction of India is particularly inaccurate. It is shown without a peninsula, suggesting no real geographical knowledge, despite an inscription noting that the Portuguese under Vasco da Gama had reached the region in 1498. The geography of the islands of the Indian Ocean is similarly confused, and the names given throughout the region bear little connection to those provided by their Portuguese discoverers.

► GOG AND MAGOG In the top right corner, at the furthest limits of the map's coverage of the geography of northern Asia, de la Cosa reproduces the fantastical topography of the *Catalan Atlas* (see pp.62–65). Here the monsters Gog and Magog are shown, one half-dog and the other with his head in his chest, apparently eating human flesh. Their proximity to a compass rose emphasises the contradictory scientific and mythological beliefs that shaped the making of this map.

8

7

Map of Venice

1500 ▪ WOODCUT ▪ 1.34M × 2.81M (4FT 4¾IN × 9FT 2½IN) ▪ MUSEO CORRER, VENICE, ITALY

JACOPO DE' BARBARI

SCALE

Jacopo de' Barbari's map of Venice is a landmark achievement in the history of both cartography and printmaking. It was the first bird's-eye view of a city, produced using wood blocks to print the six large sheets that comprise the image. The map's aerial view shows Venice with its fish-like outline seen from the southwest, with the outlying islands of Murano to the north, and Torcello, Burano, and Mazzorbo to the northeast, in the distance. The mainland (labelled *terraferma*) and the islands

of Giudecca and San Giorgio Maggiore make up the foreground, with the Alps visible in the distance along the top of the map.

An idealized view

This map is the result of a unique collaboration between two itinerant craftsmen: the German publisher Anton Kolb, and the artist responsible for its beautiful design, Jacopo de' Barbari. Unusually, in 1500, Kolb asked the Venetian authorities to give him sole publication rights and exemption from export duties on the map. He clearly believed it was unique, and at a time when printers routinely stole each other's designs, he wanted to ensure such an expensive object made a profit and was not copied.

The map's most striking feature is its scrupulous attention to realistic detail, right down to the smallest church and piazza. Even today, visitors to Venice could make their way around the city using it for reference. Debate still rages as to just how Jacopo surveyed the city in such detail, and how he transformed his findings into such a brilliantly executed perspective view. The map presents an ideal image of Venice as its rulers wished it to look, seen from above as if by a bird – or a god.

> The artist has a strangely acute eye for the real… in this hypnotic masterpiece that is half-landscape, half-cartography "

JONATHAN JONES, BRITISH ART CRITIC

JACOPO DE' **BARBARI**

c.1460–c.1516

An Italian painter and engraver, Jacopo de' Barbari is best known for his distinctive *trompe l'oeil* paintings, which are rendered in a highly meticulous, realistic style.

Born in Venice, Jacopo acquired his reputation while working at the Nuremberg court of Maximilian I, the Holy Roman Emperor, where he worked with, and learned from, the great German painter Albrecht Dürer. He moved around the great courts of Germany, where he became known as Jacop Walch – a surname that was probably derived from the German word "*wälsch*", meaning "foreigner". In fact, the surname "Barbari" may only have been adopted for the first time when he returned home to Italy, where Germans were still regarded as somewhat "barbaric". The realist approach to painting that prevailed north of the Alps may have inspired Jacopo to create his famous series of exquisite still life paintings and portraits. These demonstrate an eye and temperament for painstaking detail – the key characteristics of this monumental map of Venice.

Visual tour

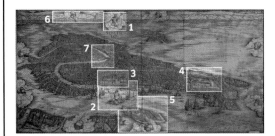

KEY

1

2

▲ **MERCURY** Perched on a cloud, Mercury, the patron of commerce (a subject close to Venice's heart) looks down protectively on the city. He sits above the caption that translates as, "I Mercury shine favourably on this above all other emporia". He also holds a caduceus (staff), which was also the symbol Jacopo used to sign many of his pictures, making this both a marker of the god and de' Barbari's signature.

▲ **NEPTUNE** Across from St Mark's Square, Neptune, lord of the seas, looks towards Mercury and announces, "I Neptune reside here, smoothing the waters at this port". Venice saw itself as Neptune's bride, offering gifts to the god to assuage the unpredictable quality of the seas, upon which it relied for its fame and fortune. As with Mercury, Neptune is an allegorical representation of Venice's desire for mastery over trade and the oceans.

▶ **SAN MARCO, DUCAL PALACE** At the centre of the map is Jacopo's minute depiction of the symbolic and political heart of Venice, St Mark's Square and the Ducal Palace. There is one particularly telling detail: following a fire in 1489, the roof of the bell tower in St Mark's Square was temporarily flat, as shown here. Later maps show the tower's newer sloped roof, as the printed blocks were subsequently altered.

3

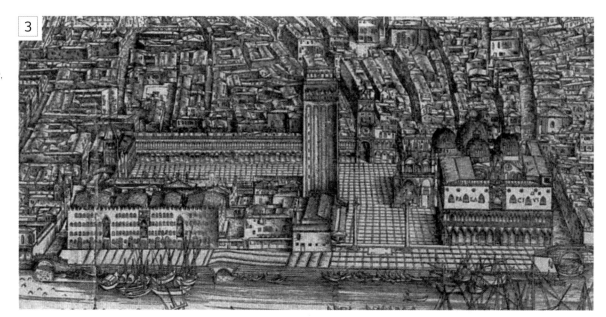

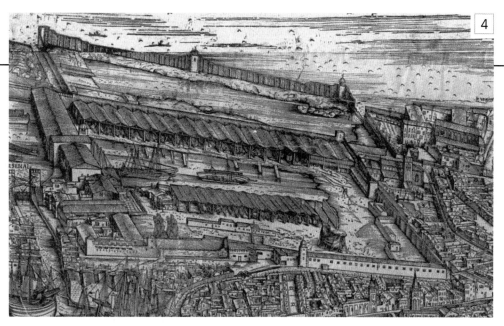

◀ **ARSENALE** Another great Venetian structure, the Arsenale, was the hub of Venice's maritime power. This vast complex included state-owned shipyards for highly efficient building of both commercial and naval ships. At its peak, it could produce a ship per day. The complex also included armouries for the production and storage of firearms.

◀ **SAN GIORGIO** Almost directly opposite St Mark's Square is the island of San Giorgio, home to a Benedictine monastery since 982 CE, and its campanile, built in 1467. Work on Andrea Palladio's famous church, San Giorgio Maggiore, which now dominates the foreshore, only started in 1566.

ON **TECHNIQUE**

Jacopo's fascination with geometry and measurement can be seen throughout his work (including his portrait of Luca Pacioli, below), and reaches an apex in the obsessive attention to detail shown in this map. Late medieval methods of ascertaining distances and orientation included pacing streets and buildings, or using rods and cords to measure them. For a labyrinthine city like Venice, such methods were of limited use. Instead, Jacopo seems to have used simple trigonometry to calculate angles and distances and then to create a basic plot, before adding the buildings and foreshortening the view to create the map's dramatically oblique perspective. Despite his success, these methods were not adopted by surveyors until centuries later.

▲ **WIND HEAD AND MOUNTAINS** In the far distance, Jacopo mixes a new level of realism with established medieval cosmographical beliefs. The wind heads personify the different winds and their directions, while the Alps stretch out in the distance, putting Venice into a wider Italian geopolitical setting.

▶ **RIALTO BRIDGE** As if to emphasize Venice as a trading centre, Jacopo has drawn the iconic Rialto Bridge, with its famous rows of shops, set within the thriving commercial activity of the Grand Canal. The map shows a timber bridge that preceded the current stone version, which was completed in 1591.

▲ **This portrait of Luca Pacioli**, attributed to Jacopo de' Barbari, shows the mathematician and friar with his instruments and notebooks.

Map of Imola

1502 ▪ PEN AND CHALK ON PAPER ▪ 44CM × 62CM (1FT 5½IN × 2FT ½IN) ▪ ROYAL COLLECTION, WINDSOR, UK

LEONARDO DA VINCI

SCALE

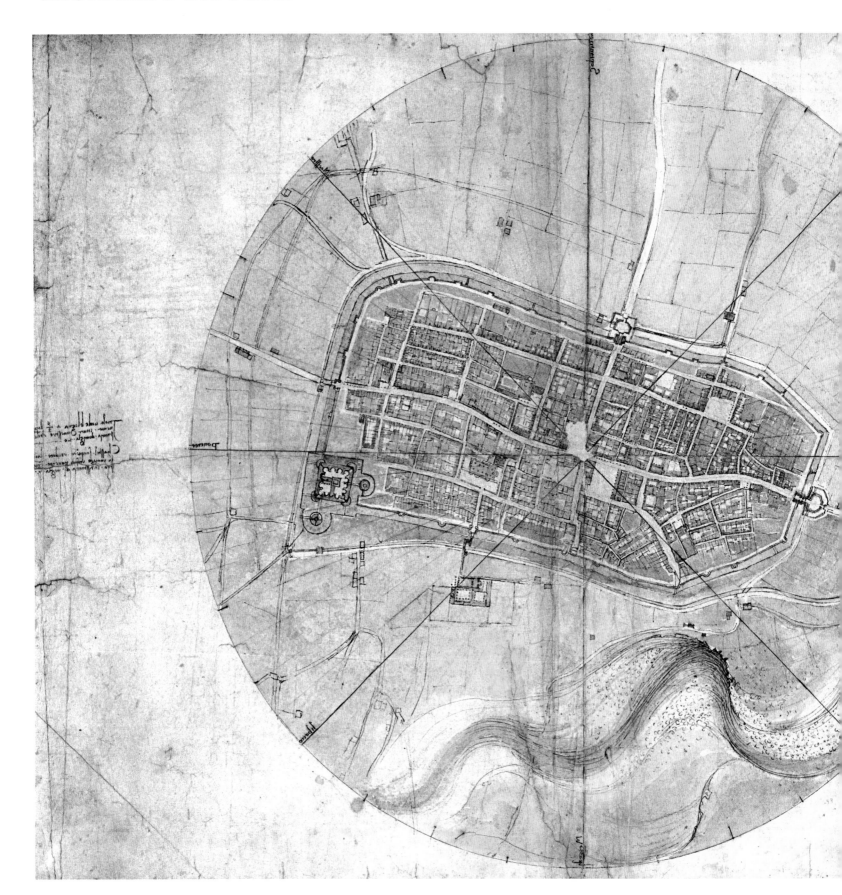

The infamous Italian military adventurer Cesare Borgia appointed Leonardo da Vinci as his architect in 1502, and charged him with the task of surveying and improving fortifications throughout northern Italy. This resulted in one of the finest maps of the Italian Renaissance. Leonardo completed his map of Imola, near Bologna, over a period of three months. In a profound departure from earlier medieval town plans, instead of adopting an oblique perspective and showing buildings in elevation (usually in order of importance rather than actual size), Leonardo chose to use an ichnographic plan – that is, one that looks directly down on the town from above, creating an almost abstract architectural plan, on which every location is perpendicular to the Earth's surface. This improved the inaccuracies of earlier maps, giving Cesare unique advantages in defending the town from attack.

LEONARDO **DA VINCI**

1452–1519

One of the greatest Renaissance artists, Leonardo da Vinci was also celebrated as an architect, musician, anatomist, engineer, and scientist.

Although his life's work spanned a huge variety of disciplines including science, technology, and mathematics, da Vinci was also responsible for some of the most iconic paintings and drawings in the world, including the *Mona Lisa* and *The Last Supper*. His obsession with representing nature and uniting art with science inevitably drew him to the field of cartography, and he made maps of various Italian regions, primarily as a military engineer in the pay of the powerful Borgia dynasty. Da Vinci also used his work in mapmaking as an occasion to experiment with different methods of surveying territory, and of representing it on a plane surface.

Old and new

Leonardo appears to have combined traditional mapmaking techniques, such as pacing out roads and squares, with newer methods, such as the use of a simple theodolite (an instrument that measured the angles between places). He does not seem to have used a compass, instead allowing himself some artistic licence in redesigning some of the city's streets. Nevertheless, modern aerial photography has confirmed the remarkable accuracy of Leonardo's achievement.

Visual tour

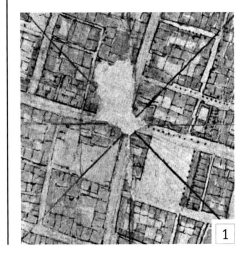

KEY

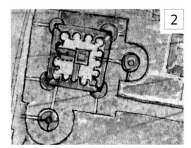

◄ **FORTIFICATIONS** There is no attempt to celebrate the architecture of one of the town's key fortifications – instead, this is an objective assessment of position and relative strengths and weaknesses for repelling enemy attack. Buildings are all seen in aerial plan.

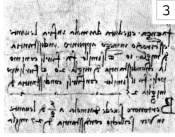

◄ **LEONARDO'S NOTES** Leonardo's notes are written in his trademark "backward" or "mirror" style, and describe defence and fortifications. The note to the left starts: "Imola sees Bologna at five-eighths from the Ponente towards the Maestro at a distance of twenty miles".

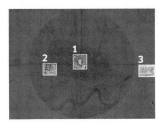

◄ **RADIAL LINES** The whole city is drawn according to a circle divided into eight wind directions (rather than compass directions), each broken down into eight further segments, with a total of 64. All lines converge on the town's centre, from which all its defences would be organized.

First Map of America

1507 ▪ WOODCUT ▪ 1.28M × 2.33M (4FT 2¼IN × 7FT 7¾IN) ▪ LIBRARY OF CONGRESS, WASHINGTON, DC, USA

SCALE

MARTIN WALDSEEMÜLLER

Although not necessarily the most famous map in this book, this is certainly the most expensive. In 2003, the US Library of Congress paid a German aristocrat a record-breaking US$10 million (£6.1 million) for what has become known as "America's birth certificate", as it is the earliest map to show and name America as a separate continent. It was made in Saint-Dié, a small town in what is now northeast France, by a team of humanist scholars. One of these, Martin Waldseemüller, is thought to have designed the map, while others wrote, translated, and printed it.

The discovery of the New World

This work is entitled "A Map of the World according to the tradition of Ptolemy and the voyages of Amerigo Vespucci and others". It is a remarkable synthesis of the latest information that had made its way back to Europe during the "Age of Discovery", when explorers ventured further east and west, expanding the limits of the known Western world. Waldseemüller was hugely influenced by the voyages of the Florentine adventurer Amerigo Vespucci (shown in the top right of the map), who claimed to have undertaken a series of journeys to the New World between 1497 and 1504. Vespucci refuted Columbus's belief that

the newly discovered lands were part of Asia, arguing instead for the existence of an entirely new continent, which eventually took his name – America. Alongside contemporary reports from explorers, Waldseemüller

MARTIN **WALDSEEMÜLLER**

c.1445–c.1521

Born near Freiburg in the Holy Roman Empire (now in modern-day Germany), Waldseemüller was the son of a butcher. He initially trained to enter holy orders, but went on to become a brilliant theology scholar instead.

He worked first in the printing business before entering the service of René II, Duke of Lorraine, and joining his humanist circle in Saint-Dié. Here he first began work on this map, which he published alongside a manual on geography and a terrestrial globe. Following its publication, Waldseemüller published more maps but he never again used the name "America", probably having been convinced that Vespucci's claims to have discovered a separate continent were not as strong as he first believed. He stayed in Saint-Dié for the rest of his life, making maps and working as a canon at the town's church until his death in around 1521.

and his team used the classical geography of Ptolemy (see pp.24–27), who can also be seen at the map's top left. The influence of Ptolemy's second projection can be discerned in the map's strange, bulbous shape, which strains to incorporate the new geographical discoveries. Much of the eastern hemisphere copies Ptolemy, but to the west the world map has been completely redrawn.

The act of printing such a large, wall-hung map on 12 sheets of paper was an enormous undertaking, and this is one of the finest examples of woodcut printing of its time. Many mysteries still surround it: how did Waldseemüller know about the Pacific Ocean seven years

The birth document of America

PHILIP D. BURDEN, BRITISH MAP DEALER AND AUTHOR

before its official discovery? Why is America so distorted? Why did Waldseemüller abandon the term "America" in any subsequent maps? And why did this map disappear for several centuries?

Visual tour

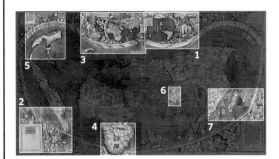

KEY

▶ **VESPUCCI**
The Florentine explorer Amerigo Vespucci stands next to an inset map of his new discovery, which also displays the complete view of the same region seen on the large map (although the two versions of the Americas are significantly different). Vespucci holds a pair of compasses, displaying his scientific credentials as an explorer.

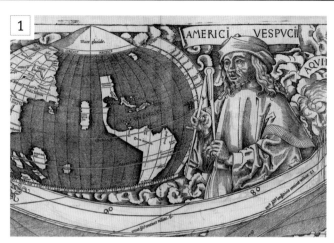

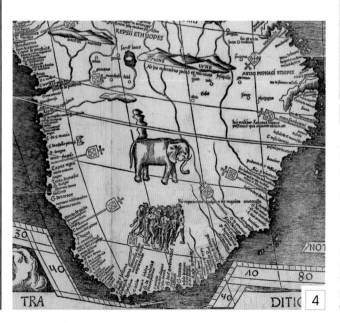

▲ **AMERICA EMERGES** Near what is now Argentina, Waldseemüller uses the word "America" for the first time on any map. He used a mixture of Spanish and Portuguese sources for his map, so his coastline, place names, and captions reflect their rivalry over the area. However, the map was dedicated to a relative of the Spanish emperor, and includes references to Columbus, suggesting that Waldseemüller ultimately supported Spanish dominion.

▲ **PTOLEMY** In contrast to Vespucci's dominance over the new continent, a portrait of Ptolemy (the "Alexandrian cosmographer") gazes down over the Old World. It shows the father of geography holding a quadrant, with the classical *ecumene* (inhabited world) centred on the Mediterranean, from where he originated. It is as though both Ptolemy and Vespucci are battling over how the Earth should be drawn.

◀ **AFRICA BREAKING THE FRAME** Portuguese flags mark the nation's recent discoveries in Africa, including the rounding of the Cape of Good Hope in 1488. This event shattered Ptolemaic geography, represented here by the way the map dramatically extends beyond the frame of Ptolemy's second projection.

5

◄ **NORTH AMERICA AND THE PACIFIC COAST** In response to recent European discoveries, Waldseemüller stretches the southern and northern latitudes of the known world described by Ptolemy, creating serious distortion at the map's limits. This is particularly evident in North America and its Pacific coastline, which seems radically foreshortened and ends abruptly in a mountain range. Erroneous mathematics appear to have been the cause of such topographically impossible coastlines, rather than actual geographical observations.

IN **CONTEXT**

Waldseemüller's map was first published in a print run of 1,000 copies, but was soon overtaken by newer, more accurate maps, and gradually disappeared. No copies were believed to have survived until 1901, when a Jesuit clergyman, Father Joseph Fischer, discovered the last remaining one, alongside its geographical manual and globe designs, in the library of the Renaissance castle Schloss Wolfegg in southern Germany. Fischer's amazing discovery led to an attempt by the US Library of Congress to buy the map, which finally reached fruition in 2003, after much careful political manoeuvring between Germany and the United States.

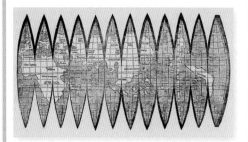

▲ **Besides wall maps**, Waldseemüller also made gores (curved segments) for creating globes. These were sold along with the world map in 1507.

◄ **INDIA** While this map improved geographical knowledge of the western hemisphere, its view of the east largely reproduced Ptolemy's vague beliefs. India is shown here without any recognizable peninsula, although the Portuguese Vasco da Gama's arrival in Calicut in 1498 is recorded. The crescent moon acknowledges the presence of Islam in the Indian Ocean.

6

▶ **DISTORTION IN THE FAR EAST** Waldseemüller's representation of the east also suffers some distortion. This includes the Indonesian archipelago and the spice-producing islands of the Moluccas, which were just beginning to attract European attention. Again, he barely updates Ptolemy's geographical knowledge here, showing "Java Major" and a group of hazily understood islands east of a completely fictitious peninsula.

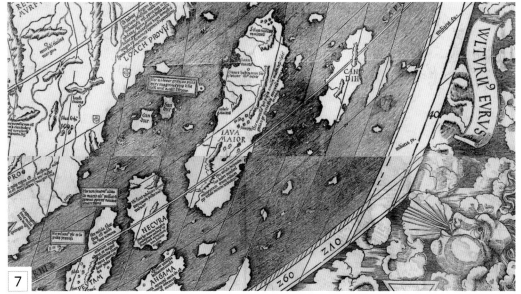

7

Piri Re'is Map

1513 ▪ PARCHMENT ▪ 90CM × 63CM (2FT 11½IN × 2FT 3IN) ▪ TOPKAPI SARAYI MÜZESI, ISTANBUL, TURKEY

SCALE

PIRI RE'IS

One of the earliest maps to be made of the Americas following their discovery by Christopher Columbus, the *Piri Re'is Map* was made not by a Spaniard, Portuguese, or Italian – the foremost seagoing nations of the day – but by a Turkish naval captain. It has fascinated generations of historians, for a variety of reasons. The original included the entire world as known by the Ottomans in the 16th century, but only the western third of the map survives today. It shows the Iberian discoveries in the Atlantic and Caribbean - as well as the landfalls made along the northern and southern coasts of the Americas – and is significant as the only 16th-century map to depict the New World in its correct longitudinal position in relation to the African coast.

Navigation and discovery

The map is a type of portolan sailing chart (see pp.68–71) – featuring a cartographic technique that records estimated distances and compass directions between ports – and is drawn on gazelle skin parchment, with compass roses and scale bars providing navigational assistance. The copious notes and 117 place names, written in Ottoman Turkish, provide a wealth of information on the circulation of geographical knowledge between Christian Europe and the Islamic Ottoman Empire. The annotation written over the lower left portion of the map explains that 20 maps were used, including those of Alexander the Great, Ptolemy, "four maps recently made by the Portuguese", and "a map of

the western parts drawn by Columbus". At the bottom of the map runs a coastline that has been interpreted fancifully by some as depicting Antarctica. However, the more interesting questions concern what the rest of the map might have looked like, and how Piri Re'is obtained such precise information on Spain and Portugal's latest discoveries, which were kept as closely guarded secrets.

PIRI **RE'IS**

c.1465–c.1553

Piri Re'is made a major contribution to cartography with his 1521 book *Kitab-ı Bahriye* ("Book of Navigation").

Despite his status as a leading Turkish historical figure, little is known about the early years of Piri Re'is. Like his famous corsair uncle, Kemal, Piri took the name "Re'is", meaning "captain", after distinguishing himself in various Ottoman naval battles and victories, including the conquests of Egypt (1517), Rhodes (1522), and Aden (1548). Piri Re'is rose to the rank of admiral in charge of the Ottoman fleet in the Indian Ocean and drew maps and navigational textbooks based on his extensive naval experience. He was executed in 1553 by the Ottoman authorities in Egypt for failing to defeat the Portuguese in the Persian Gulf.

One of the **most beautiful, most interesting, and most mysterious maps** to have survived from the **Great Age of Discoveries**

GREGORY C. MCINTOSH, AMERICAN SCHOLAR AND ENGINEER

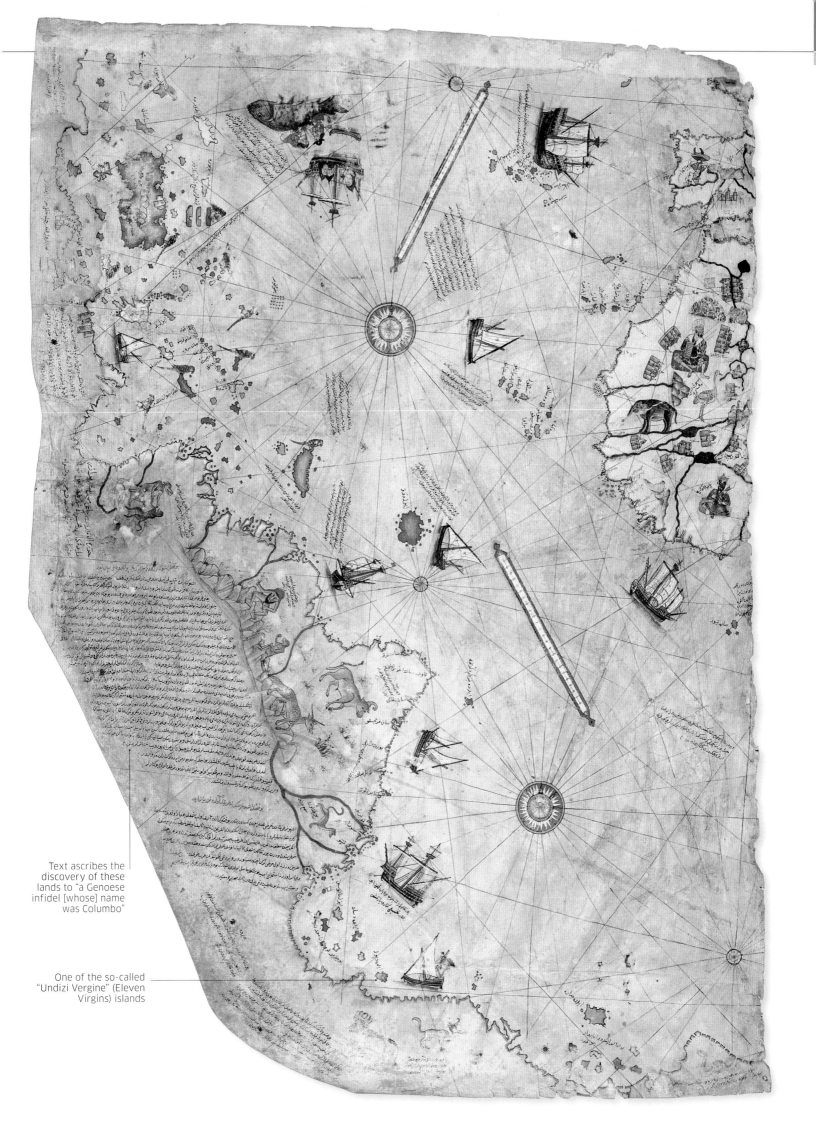

Text ascribes the discovery of these lands to "a Genoese infidel [whose] name was Columbo"

One of the so-called "Undizi Vergine" (Eleven Virgins) islands

Visual tour

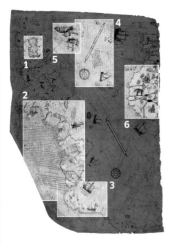

KEY

▼ **HISPANIOLA** Of all the Caribbean islands encountered by Columbus, Hispaniola is given greatest prominence on the Piri Re'is map. Shown with mountains, towns, and fortifications (as well as a parrot), it is labelled "Izle despanya" (Spanish Island) and is (to modern eyes) the wrong way up. On discovering the island, Columbus believed it was Japan (known as Cipangu), a mistake that Piri Re'is seems to have reproduced in the shape of the island.

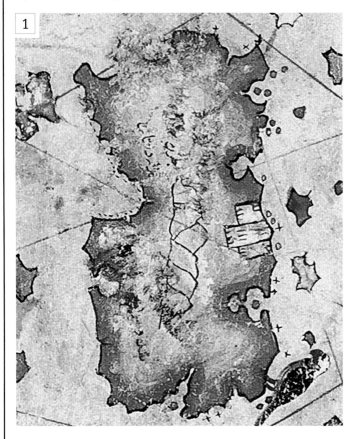

1

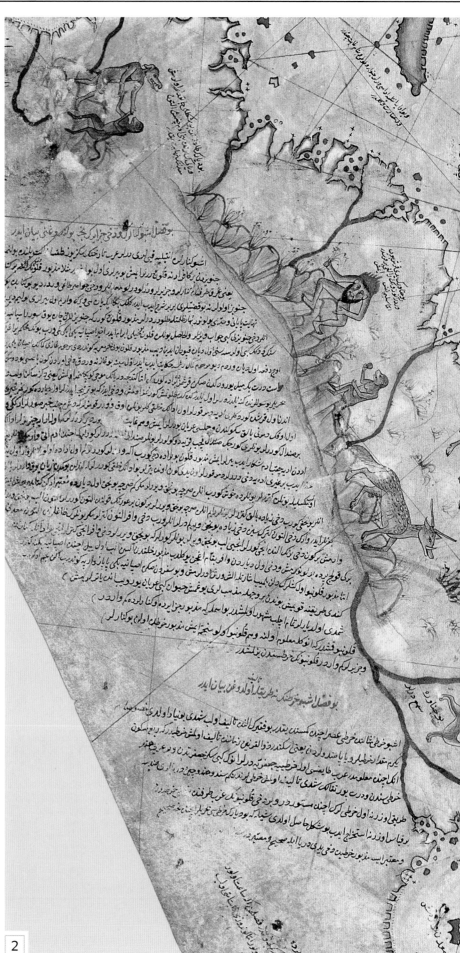

2

▶ **SOUTH AMERICA WITH MONSTROUS CREATURES** To the Ottomans, the American New World was as strange as the Far East was to medieval Christian mapmakers. Along its coast, fantastical versions of native llamas, pumas, monkeys, and even stranger creatures have been drawn, including a man with his face in his chest – remarkably similar to a race named "Blemmyes" that appear on medieval mappae mundi (see pp.56–59). The unknown interior is given over to Turkish text.

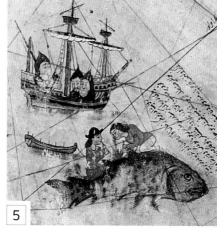

3

▲ **ANTARCTIC COASTLINE** Some of the more eccentric interpretations of the map have argued that the coastline at its foot resembles the Antarctic coast before the continent was covered in ice. They have theorized that the region was mapped by the Chinese, a lost civilization, or even extraterrestrials. The text reads: "This country is a waste. Everything is in ruin and it is said that large snakes are found here… these shores are… very hot".

IN **CONTEXT**

The Ottomans were one of the great imperial powers in the 16th-century Mediterranean, with a long and distinguished tradition of making portolan charts that fused Islamic, Christian, Greek, and Jewish knowledge. Piri Re'is produced the *Kitab-I Bahriye* ("Book of Navigation") in 1521, with beautifully detailed maps of the Mediterranean, emphasizing its shared cartographic heritage.

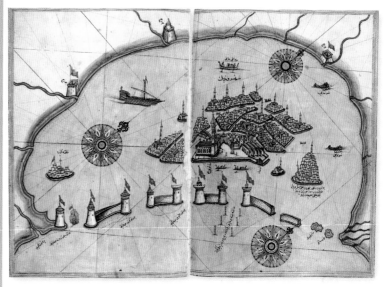

▲ **First published in the** *Kitab-I Bahriye*, this historical chart of Venice is now held at Istanbul University Library in Turkey.

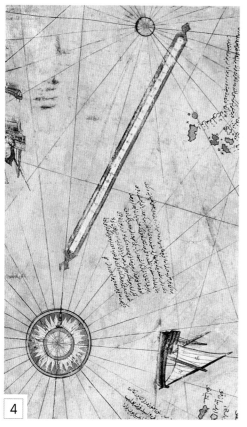

4

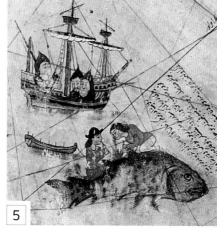

5

▲ **ST BRENDAN** The map's top corner shows the story of St Brendan, an Irish explorer and monastic saint. Its caption translates as: "He encountered this fish and, taking it for dry land, lit a fire on its back. When the back of the fish grew hot, it dived under the water." The map attributes this information as "taken from old mappae mundi", although how a Turkish mapmaker gained access to European mappae mundi remains a mystery.

▲ **COMPASS ROSE AND SCALE BAR**
As with most contemporary portolan sailing charts, the map includes a compass rose (with 32 lines showing winds and direction) and a scale bar, positioned in the north Atlantic. The rose's east–west line appears to demarcate the Tropic of Cancer. The map's scale is highly inconsistent, with the Americas shown on a much larger scale than Africa or Asia; however, this is similar to many maps of the time.

▶ **NORTHWEST AFRICA** The northwest coast of Africa was commonly known to Ottoman and Iberian mapmakers. The Portuguese had been exploring the area since the 1420s, by which time Islam was firmly established in places such as Mali. Piri Re'is shows the area's flora, fauna, towns, and rulers with great accuracy, and is almost identical to European charts of the same period.

6

Map of Utopia

1518 ▪ WOODCUT ▪ 17CM × 11CM (7IN × 4¼IN) ▪ BRITISH LIBRARY, LONDON, UK

SCALE

AMBROSIUS HOLBEIN

The writer Oscar Wilde once said that a map "that does not include Utopia is not even worth glancing at". One of the earliest and strangest examples of a utopian map is Ambrosius Holbein's 1518 woodcut. It illustrated the second edition of Thomas More's *Utopia* (first published in 1516), the text that began the tradition of utopian literature, describing a so-called perfect society. More's title was a pun incorporating the Greek words *ou* (not), *eu* (good), and *topos* (place), which combine to mean "good place", but also "nowhere" – a concept Holbein embraced. Many of the tale's characters are also contradictory: its narrator is an explorer called Raphael Hythlodaeus, whose name in Greek means "speaker of nonsense".

Holbein reproduces the island exactly as described by More: it is 320km (200 miles) wide, crescent-shaped, and has a large bay and a river running through the middle. There are, however, strange mistakes and inconsistencies, and if looked at in a certain way, the map resembles a skull, suggesting that death lurks in the midst of all utopian dreams.

AMBROSIUS **HOLBEIN**

1494–1519

Ambrosius Holbein was the elder brother of painter Hans Holbein the Younger, and trained in the new craft of printmaking.

While it is known that Holbein was born in the town of Augsburg (see pp.96–99) in what is now Germany, comparatively little is known about his career, which has been overshadowed by that of his more successful and talented brother. However, records show him working alongside Hans in Basel, Switzerland, where he also joined the local painters' guild. Only around a dozen of his paintings survive, most of them portraits. This makes Ambrosius Holbein's involvement in the woodcut map of Utopia all the more intriguing.

Visual tour

KEY

◁ THE CITY OF "AMAUROTUS" Utopia's capital city to the north is called "Amaurotus". Is More being serious when he describes it as an ideal city? Its name comes from the Greek for "shadowy" or "unknown".

▷ RAPHAEL HYTHLODAEUS The narrator of More's *Utopia*, Raphael Hythlodaeus, is named and shown in animated conversation with one of the book's other characters (possibly Peter Gillies). He points up towards the island, as if describing it to his companion.

◁ THE MOUTH OF THE ANYDRUS RIVER The map is full of jokes – or possibly mistakes. The name "Ostium anydri" means "mouth of the Anydrus river", but what is actually shown is the river's origins (a waterfall), not the point where it meets the sea.

Amaurotū vrbs.

Fons Anydri.

Ostium anydri.

Hythlodaeus.

Map of Augsburg

1521 ▪ WOODCUT ▪ 80CM × 1.91M (2FT 7½IN × 6FT 3¼IN) ▪ BRITISH LIBRARY, LONDON, UK

JÖRG SELD

SCALE

This panoramic map of Augsburg in southern Germany is the first ever printed plan of a northern European city. Published using eight wood blocks, it looks down on the city from a dramatic, oblique angle, celebrating

Augsburg's status as one of the most important cities of the Holy Roman Empire. The city regarded itself as a northern European Rome, and the map's maker, Jörg Seld, drew on Italian Renaissance developments in mathematics

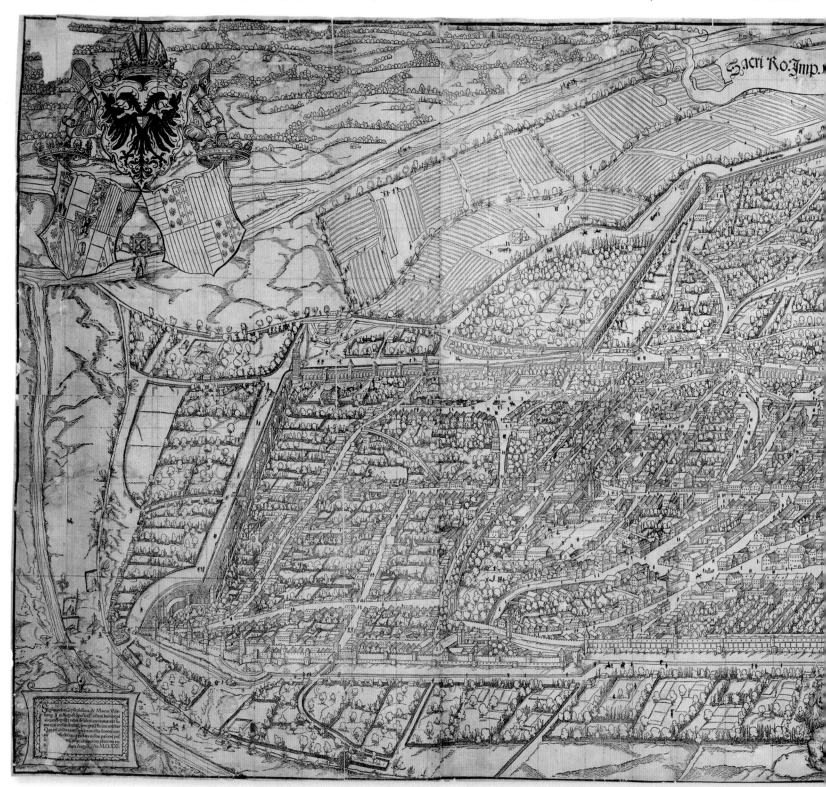

and surveying, which are described in its various panels, to render each street and building in meticulous detail. Seld had surveyed the city in 1514, but it was not until he collaborated with an artist, Hans Weiditz, and a team of local apothecary-printers that the map finally saw publication. It was also funded by the wealthy Fugger merchant dynasty who dominated the city, especially Jakob Fugger (1459–1525), the man responsible for building part of the city known as the "Fuggerei". Seld and his collaborators combined technical virtuosity with mathematical brilliance in this compelling cartographic vision of civic pride.

JÖRG **SELD**

c.1454–1527

Surprisingly little is known about the details of Seld's life considering his prominence as an artist and craftsman. He was born and died in his beloved Augsburg, where he was a pivotal figure in its artistic and civic life.

He trained as a goldsmith and became a master worker in 1478. By 1486, he was working on the city's basilica of St Ulrich and Afra, and is known for his silver altar in the cathedral, and his involvement in the shift in Augsburg's architecture from High Gothic to an Italianate Renaissance style. Also a military engineer, he designed structures throughout Germany, and developed an interest in different artistic styles, particularly Jacopo de' Barbari's civic mapmaking (see pp.80–83). Today, his reputation rests on his map of Augsburg, with its debt both to architectural design and the craft skills of cutting large wood blocks for printing.

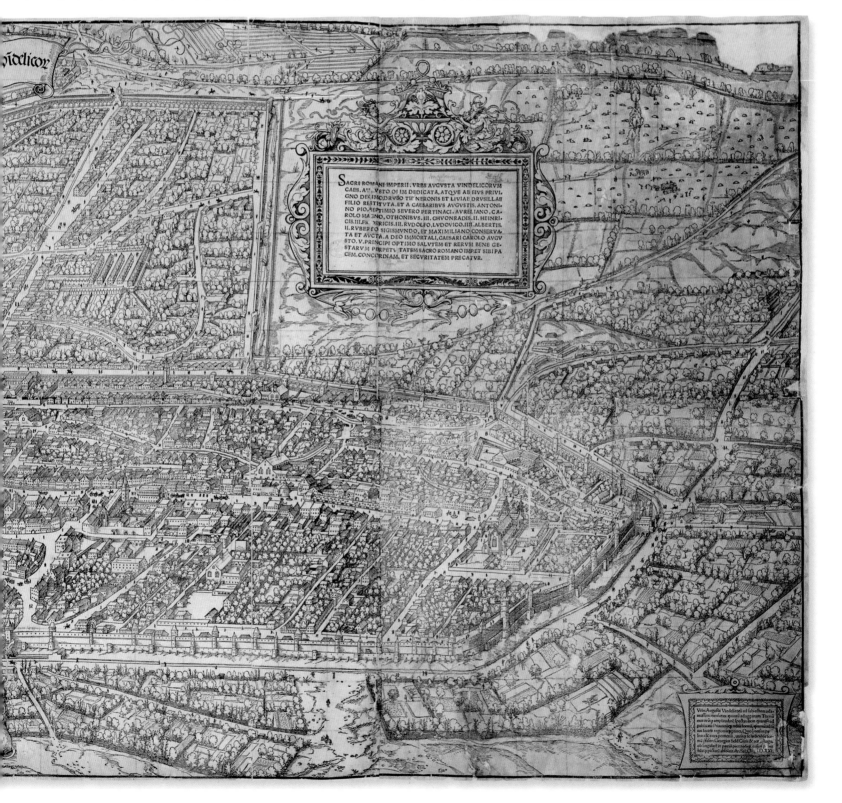

Visual tour

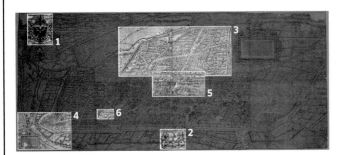

KEY

▶ **IMPERIAL EAGLE**
The Holy Roman Empire still existed nominally within Europe in the 16th century, with cities such as Augsburg ruled by the latter-day Holy Roman Emperor, the Habsburg Charles V. The map shows its allegiance in the top left, with the doubled-headed Roman eagle above the arms of Emperor Charles.

▶ **CIVIC COAT OF ARMS** The map proudly displays its civic coat of arms, a pine cone (dating back to its time as a Roman capital) flanked by rampant lions, suggesting a tension between civic pride and imperial allegiance.

▼ **THE FUGGEREI** To the north of the city, Seld depicts the Fuggerei, a town within a town, endowed by the Fuggers in 1516. It comprises subsided housing for impoverished Augsburg workers, and is still inhabited today.

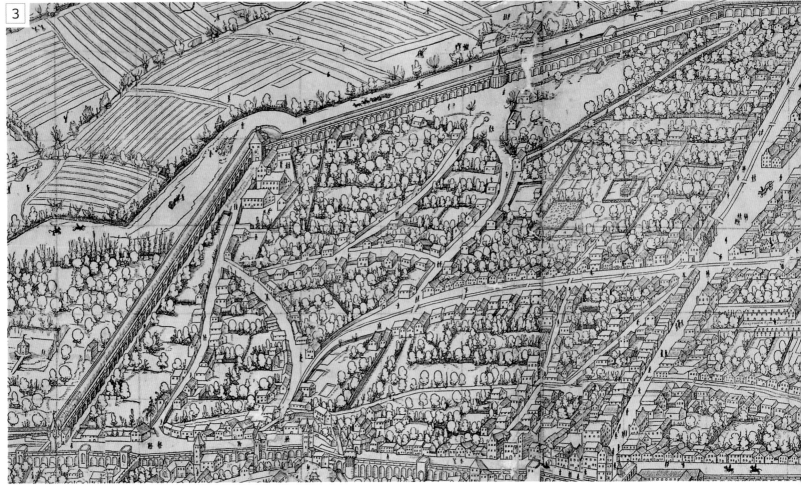

▼ BEYOND THE CITY WALLS Outside the civility of Augsburg's city walls, the land is depicted as a sparsely populated and potentially dangerous space. The panel states that the map was designed for those who either miss the city or want to know more about it, and celebrates a "singular love of one's native land".

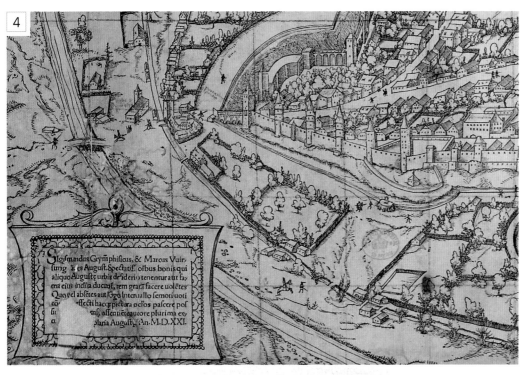

IN **CONTEXT**

Like Jacopo de' Barbari's earlier view of Venice (see pp.80–83), Seld used a variety of new surveying techniques to create a detailed image. By the 1520s, wealthy northern European cities, such as Augsburg, were developing topographic surveys for the various territories of the Holy Roman Empire, and to celebrate the civic values of burgher industry. Seld probably used his experience as a military engineer to pace the city's dimensions, using simple geometrical methods to measure distances and angles, which were then carved into the huge wood blocks by the artist Hans Weiditz prior to printing.

▲ Jacopo de' Barbari's map of Venice shows a similar level of detail to Seld's Augsburg map.

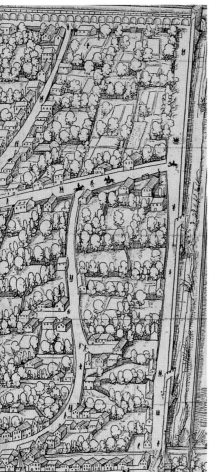

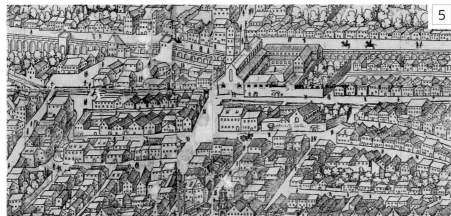

◄ THE HEART OF THE CITY
Seld represented each civic building at the city's centre with loving detail, from churches and municipal offices to houses and thoroughfares, including those he had designed himself. He even shows riders on horseback cantering through the streets.

◄ MYSTERY SCENE
The map celebrates the city and its people; however, not all of its meanings are clear. In the top right-hand corner, a dramatic scene shows a person lying on the ground, surrounded by others. Two people appear to be fighting, while another stands with arms raised. Is it some playful festivity, or a funeral? Could it be a robbery, or even a murder?

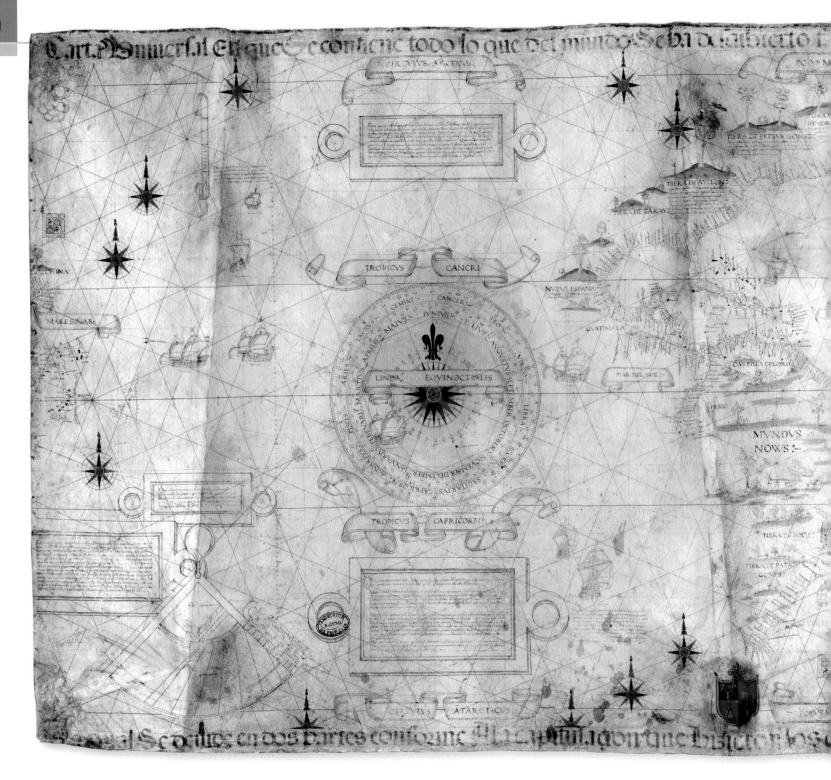

Universal Chart

1529 ▪ VELLUM ▪ 85CM × 2.05M (2FT 9½IN × 6FT 8¾IN) ▪ VATICAN LIBRARY, ROME, ITALY

DIOGO RIBEIRO

SCALE

Despite its veneer of scientific accuracy and topographical detail, this is a highly selective political map of the world. Portuguese-born navigator and cartographer Diogo Ribeiro created it to support the claim of Spain's Habsburg rulers to the spice-producing Moluccas Islands. Their claim came after the first known global circumnavigation (1519–22), led by the Portuguese explorer, Ferdinand Magellan, travelling in the service of Spain. Magellan gambled that he could reach the Moluccas by sailing southwest via Cape Horn, a quicker and shorter way than the customary eastern route. Although he perished, the surviving crew returned to claim that if the globe were divided along a

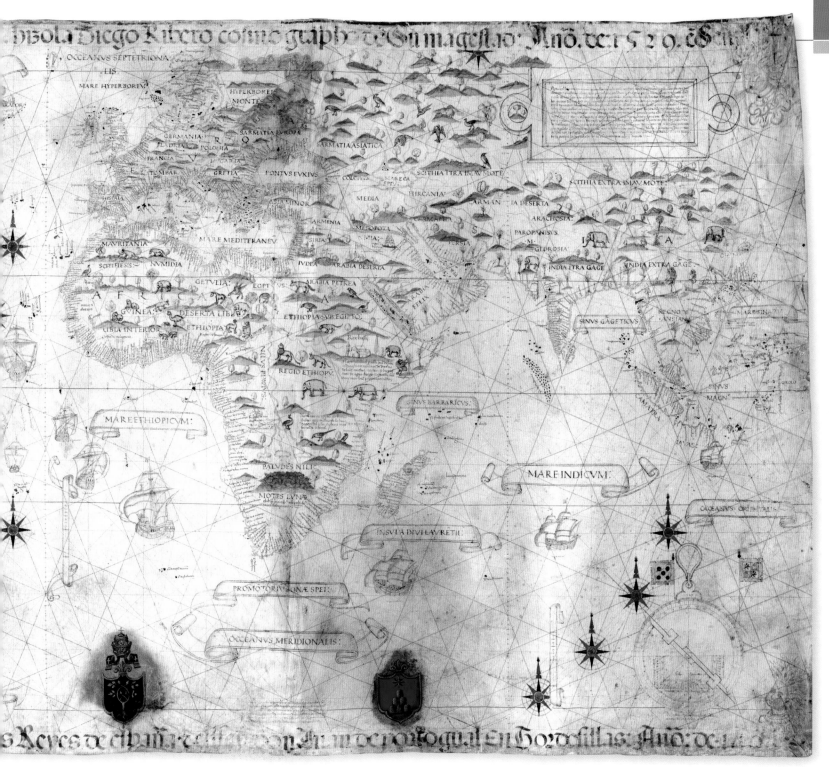

north-south line between the Spanish and Portuguese empires, the Moluccas would fall in the Spanish western half. Diplomacy tipped in Spain's favour after the Spanish paid Portuguese mapmakers such as Ribeiro to manipulate geographical reality and place the Moluccas in their hemisphere. Ribeiro did so in this Universal Chart, which "confirms" the islands' position west of a dividing meridian drawn through the Pacific.

His map is beautifully hand-drawn, full of decorative detail and the paraphernalia of the new age of scientific mapping. It also shows how Iberian mapmakers had turned their back on Ptolemy: his Old World is still there to the east, but the map's centre of gravity is the Atlantic and the relatively unknown Pacific Ocean in the west.

DIOGO **RIBEIRO**

c.1480–c.1533

Starting out as a seaman, Diogo Ribiero sailed on several voyages of exploration before settling down to work for the Spanish crown as a mapmaker and designer of scientific instruments.

Ribeiro went to sea at an early age, sailing to India with the famous Portuguese navigator Vasco da Gama in 1502. He joined several other expeditions as ship's pilot, but by 1518 he was in Spanish pay, producing maps for Magellan's voyage. In 1523, Ribeiro was named Royal Cosmographer at the Casa de Contratación ("House of Trade") in Seville, the centre of Spain's administration of its overseas empire. In 1524, he was part of the Spanish entourage debating rights to the Moluccas, and he began a series of monumental world maps. Following his success, he continued to work for the Spanish Crown, inventing various scientific instruments before his death.

Visual tour

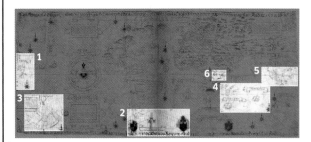

KEY

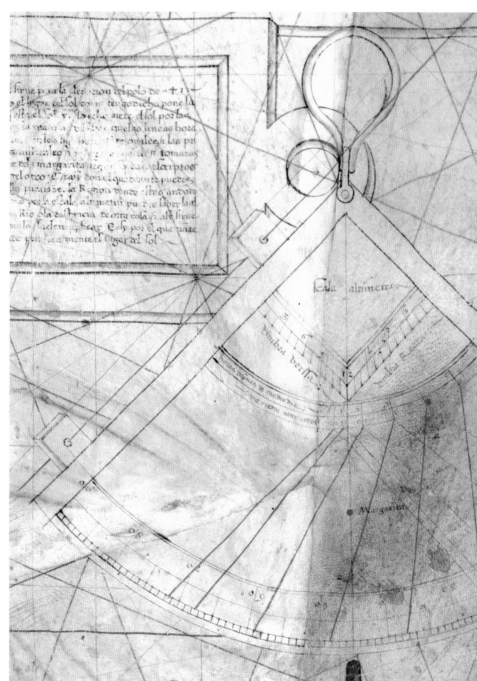

▼ **MOLUCCAS** Ribeiro drew the Molucca Islands in the far west, where they run diagonally west to east above the compass rose. By placing them just over seven degrees to the east of the Pacific dividing line, the map puts them within the Spanish half of the globe.

1

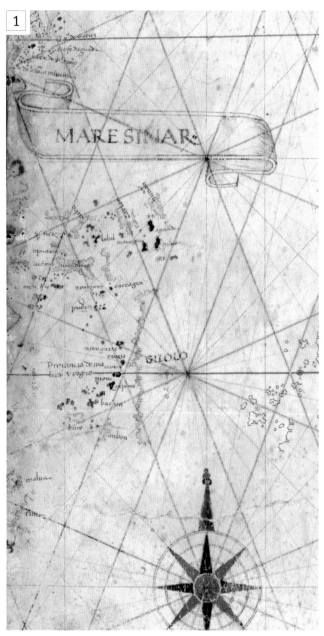

MARE SINAR

➤ **FLAGS AND COATS OF ARMS** Across the bottom of the map are a series of beautifully rendered royal flags and coats of arms, with Spain to the left (west) and Portugal to the right (east). They demarcate the relative spheres controlled by each imperial rival.

2

3 ◀ **SCIENTIFIC INSTRUMENTS**
Drawings of complicated navigational instruments such as this quadrant give the map the appearance of objectivity and authority. However, this apparent accuracy is highly selective.

▼ **SHIPS IN THE INDIAN OCEAN**
The combined commercial and political importance of monopolizing the spice trade is captured by the merchant ships depicted plying their trade in the Indian Ocean and beyond.

IN **CONTEXT**

Ribeiro's map was the first in a long line of political maps used to settle territorial disputes. In 1529, the Treaty of Saragossa gave the Moluccas to the Spanish, not the Portuguese, and both sides agreed to sign a map, much like this one, showing the islands in a mutually agreed location. Maps became legal documents binding rivals together in agreements based on geography.

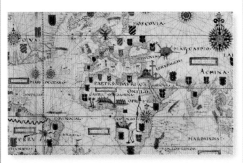

▲ **Domingo Teixeira's world map**, showing the dividing line between Portuguese and Spanish spheres of control in 1573.

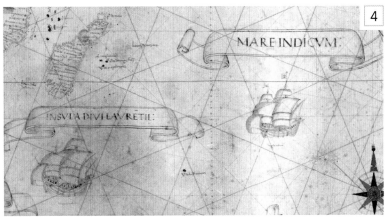

4

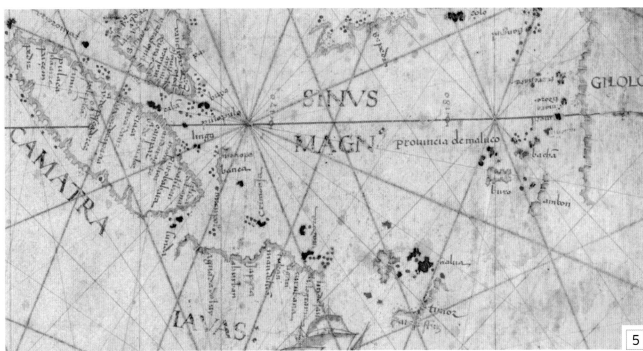

5

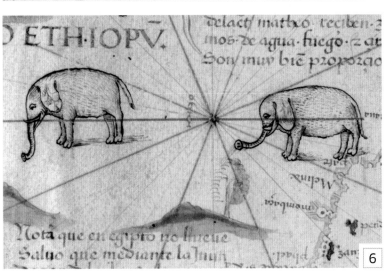

6

▲ **INDONESIAN ARCHIPELAGO** The interest in spices transformed European maps of the Indonesian region. The fantastical geography of Ptolemy and medieval mapmakers has gone, replaced with an inquisitive commercial eye for new markets. Java and Sumatra are shown here bisected by the Equator.

◀ **ELEPHANTS IN CENTRAL AFRICA** Away from the seas and coasts, Ribeiro's geographical confidence waned. In the interiors of Africa and the Americas, he recycled older fantastic geographical beliefs, putting elephants in central Africa because he clearly had little information about what was there.

Aztec Map of Tenochtitlan

1542 ▪ INK ON PAPER ▪ 32CM × 22CM (1FT ¾IN × 8¾IN) ▪ BODLEIAN LIBRARY, OXFORD, UK

UNKNOWN

SCALE

Human societies map their environments in different ways, as shown by this map of Tenochtitlan, the largest city in the pre-Columbian Americas and capital of the Aztec Empire, situated in modern-day Mexico City. It was drawn after the city fell to the Spanish conquistador Hernán Cortés in 1521, and formed part of the *Codex Mendoza*, a book detailing Aztec history and daily life.

Tenochtitlan was built on a lake – reduced here to a blue rectangular border – with four of the city's many canals forming a blue cross to divide it into four triangular districts. Tenochtitlan's residents would have recognized the city's four main areas of Atzaqualco, Teopan, Moyotlan, and Cuepopan; the 10 founding figures pictured in the quadrants; and the city's symbol of an eagle perched on a cactus (which is still Mexico's national symbol today). Key moments in the founding and subsequent history of the city appear in the bottom panel. Each of the surrounding hieroglyphs represents a year, beginning with Tenochtitlan's foundation in 1325, represented by a house crowned by two dots in the upper-left corner. The map is a unique document that represents the history and geography of the Aztecs' sacred capital, and affords a glimpse into a lost civilization just as the Spanish Empire was bringing it to a violent and bloody end.

Visual tour

KEY

▶ **EAGLE** In Aztec mythology, the god Huitzilopochtli instructed his people to settle where they saw an eagle holding a snake perched on a cactus. Below is Huitzilopochtli's symbol, a shield, and arrows – an allusion to the warlike methods the Aztecs used to win power. This vision of Tenochtitlan's foundation lies at the map's centre.

1

2

◀ **TENOCH** Left of centre and closest to the eagle is Tenoch, leader of the Aztecs and one of the city's founders. He gave Tenochtitlan its name – *tetl* means "stone", while *nochtli* means "prickly pear cactus". He sits on a woven mat and is immediately recognizable by his headdress. Debate continues today as to whether Tenoch was a real or mythical figure.

▼ **CULHUACAN** The bottom section of the map shows armed Aztec warriors in battle, defeating the rival cities of Culhuacan and Tenayuca, which are shown toppling to the ground. The map is therefore a celebration of the military foundation of the Aztec Empire.

IN **CONTEXT**

Mesoamerica was home to a rich tradition of mapmaking, much of which was ended by Spanish rule. What remains is characterized by the use of hieroglyphs to depict events and places, such as the five houses in the upper-right corner of this map, in what has been called the "spatialization of time". These maps saw space connected to time, often blending geography with history.

▲ **Aztec warriors consulting a map** in the *Codex Florentine* (c.1570), which documented Aztec culture, world view, rituals, and history.

3

New France

1556 ▪ WOODCUT ▪ 28CM × 38CM (11IN × 1FT 3IN)
▪ MEMORIAL UNIVERSITY, NEWFOUNDLAND, CANADA

SCALE

GIACOMO GASTALDI

The earliest printed map to show the region of northeast Canada as well as New York harbour in detail, Giacomo Gastaldi's map was also the first to use the title "New France" to describe French claims to the area around Nova Scotia and Newfoundland, all the way to Manhattan. It was first published in Giovanni Battista Ramusio's travel narrative *Navigations and Voyages*, and is based on the discoveries made by the Florentine explorer Giovanni da Verrazzano, who in 1524 was commissioned by the French court of King Francis I to sail along the Newfoundland coast, down as far as the bays of New York and Narragansett. Verrazzano named the region "Francesca" in honour of his patron, but it was Gastaldi's widely distributed map in Ramusio's encyclopaedic travel compendium that gave the name "New France" its authority.

Illustrative and objective

The map is as interested in telling a story about the landscape and people of the area as it is in producing an objective map of the new discovery. Ships and fishermen move across a sea filled with fish, while the mainland is populated by local Native American communities hunting, fishing, and relaxing amidst an abundance of flora and fauna. To the west, the blank "Parte Incognita" suggests this is still a world coming into European focus.

GIACOMO **GASTALDI**

1500-1566

One of the greatest Italian mapmakers of the 16th century, Giacomo Gastaldi was also a distinguished astronomer and engineer who worked for the Venetian state.

One of Gastaldi's earliest and most important works was a 1548 pocket-sized Italian edition of Ptolemy's *Geography* using copperplate engravings, which included new maps of the Americas. He had a long and distinguished career and was the first mapmaker to propose a strait separating America and Asia. He also made fresco maps for the Doge's Palace in Venice, designed the *New Description of Asia* (1574), and worked closely with Giovanni Battista Ramusio (1485-1557) on his monumental collection of travel narratives, titled *Navigations and Voyages* (1550-59).

TRAMONTANA   TERRA DE LABORADOR

A

ISOLA DE DEMONI.

TERRA NVOVA.

colfo di castelli

LEVANTE

Mote de trigo

VMBEGA

Port Réal   Port du Refuge

C. Breton

C. de breton

Bonne uiste.

Bacalaos.

Brion.

C. de ras   C. de speraza

Isola   de Bretoni.

Isola della rena.

vado à la terra nuoua

OSTRO

Visual tour

KEY

▶ **ISLAND OF DEMONS**
Gastaldi's geography becomes increasingly fragmented and semi-mythical off the Newfoundland coast, showing European fears and opportunities in the same space. To the north, the "Isle of Demons" is populated by winged creatures, while inland, local Beothuk Indians are shown going about their everyday lives.

▲ **"ANGOULESME"** The peninsula of "Angoulesme" contains the site of modern-day New York City. It was first named "New Angoulême" by Verrazzano in 1524, in honour of King Francis I, who had been Count of Angoulême prior to his coronation. Subsequently named "New Amsterdam" by the Dutch, it only became "New York" in 1664.

ISOLA DE DEMONI,

◄ CAPE BRETON This island, discovered by Italian explorer John Cabot in 1497, became a fishing colony in the 1520s. Scenes of inviting fisheries here function as an advertisement, underlining the map's commercial agenda of encouraging European investment in the area.

▼ SCENES FROM LOCAL LIFE Indigenous Native Americans are shown going about their daily routine, hunting, dancing, and cooking. The drawings of Beothuk camps composed of *mamateeks* – communal houses made from wooden poles and birch bark – are particularly striking.

IN **CONTEXT**

Gastaldi was heavily influenced by the "Dieppe School", a style of French mapmaking that flourished in around 1530–70. Named after the thriving commercial port on the northeast coast of France, its mapmakers fused the region's famed miniature painting tradition with portolan sea charts (see pp.68–71) to produce strikingly decorative maps, which were hand-drawn, exquisitely coloured works of art. These were often bound in large-format atlases to be spread out and consulted by their wealthy patrons. Italians such as Gastaldi admired them because they assimilated all the latest geographical discoveries, especially those of the Portuguese, while also promoting French imperial aspirations, especially in the New World.

▲ This 16th-century French world map builds on the achievements of earlier Portuguese sea charts.

▼ THE ISLAND OF SAND Gastaldi was interested in developing a new cartographic visual vocabulary, but it did not always work: here he depicts the Grand Banks of Newfoundland as a long island of sand.

► "PARTE INCOGNITA" At the very edge of the map's unknown territory, where European knowledge runs out, Gastaldi has added a mysterious vignette showing what appear to be two fur trappers and a bear.

A New and Enlarged Description of the Earth

1569 ▪ COPPERPLATE ENGRAVING ▪ 1.24M × 2.02M (4FT ¾IN × 6FT 7½IN)
▪ MARITIEM MUSEUM, ROTTERDAM, THE NETHERLANDS

GERARD MERCATOR

Made by probably the greatest – and certainly the most famous – of all mapmakers, Gerard Mercator, this world map is based on his mathematical projection of the world, which became the most successful of its kind in history. It represents the first cartographic attempt to plot a spherical Earth onto a flat piece of paper, enabling pilots to draw a straight line across the Earth that took into account its curvature. The key to achieving this lay in lengthening the parallels at higher latitudes, but its success came at a cost. Both the North and South Poles can be seen stretching to infinity, and other regions such as Canada and South America also appear far too large for their actual surface area. The map's subtitle claimed it was designed "for use in navigation", but it was on too small a scale to be used effectively by navigators – although by the late 17th century, most pilots had adapted it for larger scale maps and it became the standard projection for long-distance sailing.

Science and art

As well as an important scientific breakthrough in mapping, this engraving is also a work of art. Mercator was a renowned engraver who developed the distinctive curled, italic writing and decorative flourishes that cover the map. They are often used to adorn areas such as North America, where he clearly has little or no information about the interior. Despite Mercator's scientific advances, much of his geography remains influenced by classical and medieval conventions: the seas are filled with monsters as well as commercial vessels, the North Pole is inhabited by pygmies, and Asia is still depicted incorrectly, drawing on Ptolemy (see pp.24–27).

GERARD **MERCATOR**

1512-1594

Born into humble origins in Flanders, in what is now Belgium, Gerard Mercator was a brilliant scholar who studied at the University of Leuven, before working as a scientific instrument-maker, engraver, and mapmaker.

Despite creating a series of innovative maps of Flanders, the Holy Land, and the world, Mercator was arrested for heresy in 1554 by the Catholic Habsburg authorities because of his Lutheran beliefs, and narrowly avoided execution. He moved to Duisburg, in what is now Germany, where he spent the rest of his life working on a vast cosmography of the universe, which included a chronology of world history, the first atlas, and the world map on a projection that would ensure his immortality. He was the first mapmaker in history to use the term "atlas" to refer to a collection of maps.

When I saw that Moses's version of the Genesis of the world did not fit… I began to have doubts about the truth of all philosophers and started to investigate the secrets of nature

GERARD MERCATOR

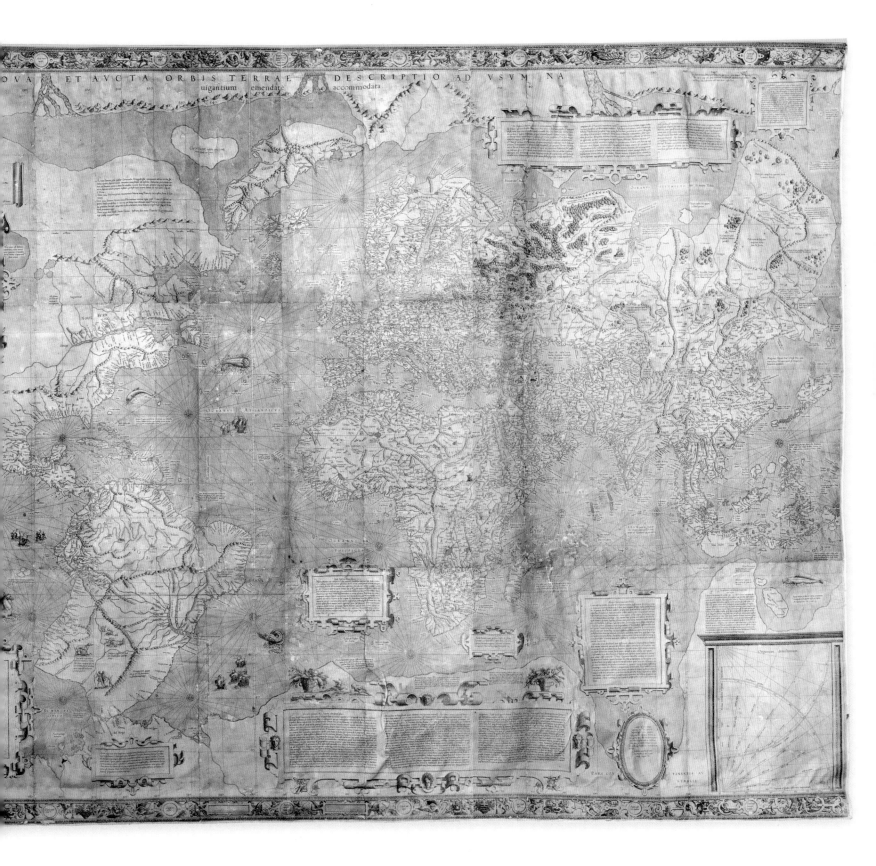

Visual tour

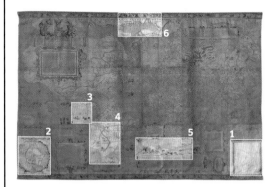

KEY

> ▶ **THE NORTH POLE WITH "INFLOWING SEAS"**
> Despite his belief in mathematical accuracy, Mercator still subscribed to some very strange and old-fashioned ideas. These included his assertion that the Earth was in fact hollow, and that the North Pole consisted of four islands with a great magnetic rock at its centre, into which the seas flowed. According to Mercator, pygmies called "Scraelings" lived on the four islands, shown here on the main map's inset.

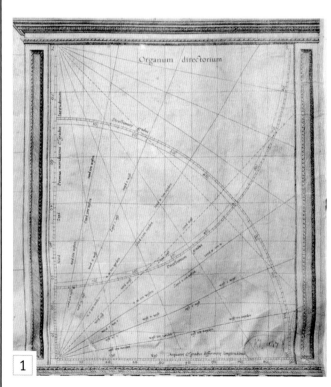

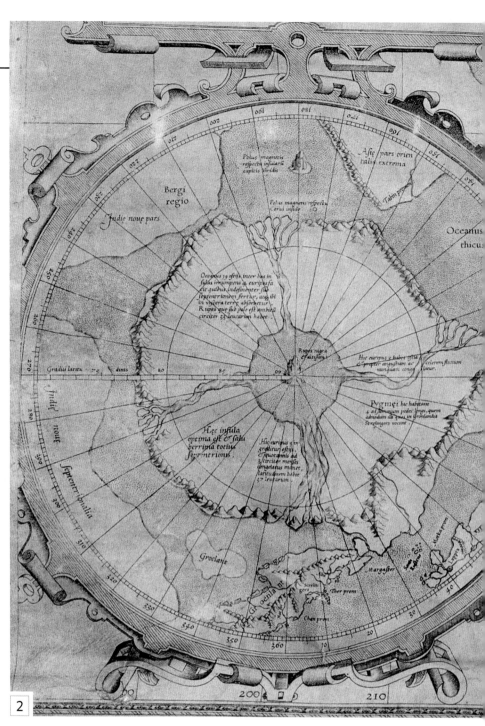

▲ **PROJECTION METHOD** To help a ship navigate along a straight course, Mercator plotted curved lines, known as rhumb lines, that crossed each meridian at the same angle. These lines mimicked the Earth's curvature and prevented the vessel from sailing off course. Mercator's solution, illustrated on this mathematical diagram, provided a method for navigating long distances that is still used today.

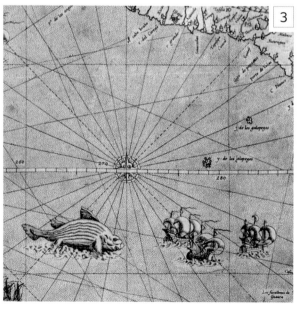

◀ **MONSTERS OF THE DEEP**
Mercator never went to sea, so he had to sift fact from fiction when reading travellers' tales, with variable results. Thus, terrifying monsters, some twice the size of Europe's largest ships, appear across the Pacific and Indian Oceans. Here, off South America, a whale-like creature bears down on a fleet of European vessels.

◄ AFTER MAGELLAN
Mercator was working in the aftermath of Magellan's first global circumnavigation, which had a profound effect on mapmakers. Here he shows South America and a circumnavigable Cape Horn, although the relative lack of information on it, when combined with the distortions caused by his projection, leaves the entire landmass looking squashed and far too close to Antarctica.

ON **TECHNIQUE**

Mercator's new projection of the world offered an ingenious solution to the classic geographical problem – dating back to the time of the ancient Greeks – of how to draw a spherical object such as the Earth onto a flat, square piece of paper without encountering significant distortion. His answer was to treat the Earth as if it was a cylinder, on which its geographical features were drawn, then unrolled, before finally lengthening the space between the parallels towards the poles, leaving the impression they go on forever. The result looks distorted north to south, but accurate east to west.

▲ **Mercator's terrestrial** globe.

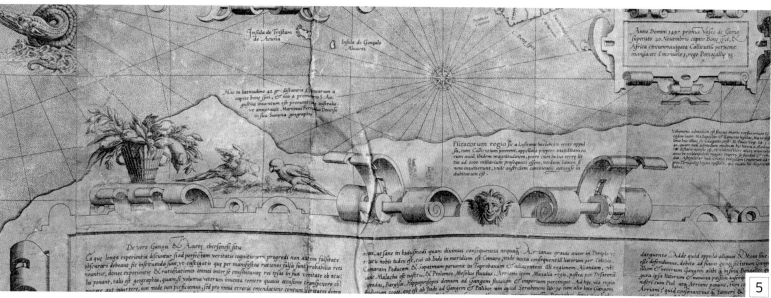

▲ TO INFINITY AND BEYOND: ANTARCTICA
Mercator's method of representing the spherical Earth as a rectangle means that the North and South Poles are so distorted that they stretch to infinity from east to west. The beautifully engraved cartouches, legends, and decorative embellishments do everything to distract the viewer's eye from the obvious distortion.

◄ A FLAT POLE The main map shows a very different North Pole to that on the circular inset (see detail 1). On the main map, the pole cannot be represented because of the projection method; all that is visible are two small islands (also shown on the inset map), preceding the four main islands, stretched along the top of the map.

NEW DIRECTIONS AND BELIEFS

- Map of Northumbria

- Vatican Gallery of Maps

- The Molucca Islands

- Map of the Ten Thousand Countries of the Earth

- The Selden Map

- Nautical Chart

- Map of the "Inhabited Quarter"

- New Map of the World

- Britannia Atlas Road Map

- Map of New England

- Corrected Map of France

- Map of the Holy Land

- Land Passage to California

- New Map of France

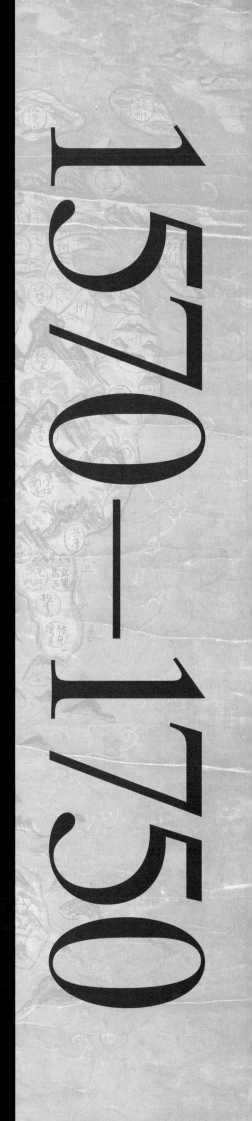

1570—1750

Map of Northumbria

1576 ▪ COPPERPLATE ENGRAVING ▪ 39.7CM × 51CM (1FT 3½IN × 1FT 8IN) ▪ BRITISH LIBRARY, LONDON, UK

SCALE

CHRISTOPHER SAXTON

In the early 1570s, estate surveyor Christopher Saxton received a commission from royal official Thomas Seckford at the behest of his superior, Queen Elizabeth I's Chief Minister, Lord Burghley. Saxton was given the task of making the first ever county maps of England and Wales. Over the next few years, Saxton mapped each county using an early form of triangulation to calculate distances and locations (see p.165).

In 1574, he published his first county map, of Norfolk, followed by others, including this map of Northumbria. Later, in 1579, he published the first ever *Atlas of the Counties of England and Wales*, which included 34 of the county maps, as well as a complete map of England and Wales. The maps were engraved and hand-coloured to a high level of detail,

with standardized symbols for topographical features such as the hills and rivers shown here. Each map also bore the stamp of authority of the royal arms (top right) and Seckford's coat of arms (lower right).

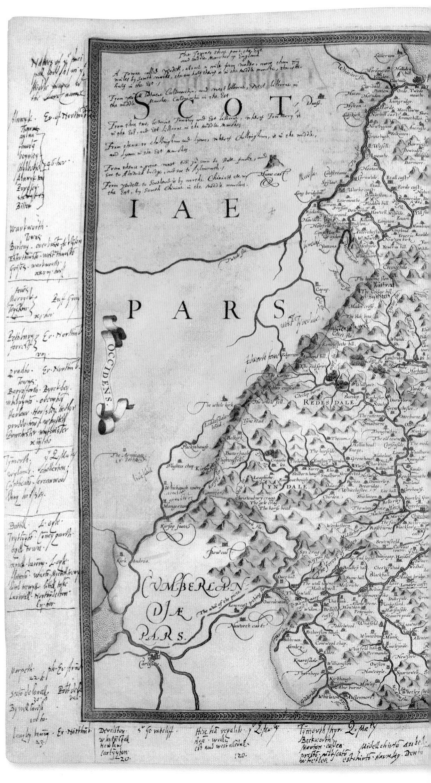

> For over **two hundred years**… nearly every printed map of **England and Wales** derived from Saxton "

RICHARD HELGERSON, *FORMS OF NATIONHOOD*

CHRISTOPHER **SAXTON**

c.1540–c.1610

Born in Yorkshire, England, Saxton was trained in estate surveying and draughtsmanship by the Vicar of Dewsbury, John Rudd, before joining the household of Thomas Seckford, an Elizabethan courtier.

Seckford bankrolled Saxton's county mapping project, probably with Lord Burghley's support. Considering the work required to survey all 34 counties of England and Wales, Saxton worked extremely quickly, completing the first map by 1574 and finishing the whole project just five years later. Debate continues as to his actual surveying methods, but it is known that the maps were designed by some of the finest engravers of the day, many of them Dutch and Flemish. Anticipating the success of his maps, Saxton successfully applied to Queen Elizabeth I for a licence granting him exclusive publication of each one for 10 years. His atlas set a benchmark for generations of future English mapmakers, and by 1580 he was firmly established as the country's leading surveyor. Future generations came to regard him as "the father of English cartography".

Saxton's maps had been an unprecedented undertaking, and they were a huge success. Burghley regarded them as crucial to national security, inspecting and annotating the first proof copy of each one: Northumbria's proximity to the border with Elizabethan England's troublesome neighbour, Scotland (shown here as blank) made this a particularly important map for the Chief Minister, who covered it with notes on local information and political intelligence.

Saxton's groundbreaking work, meanwhile, began a tradition of regional mapmaking that would culminate with the creation of the Ordnance Survey, Britain's national mapping authority, more than 200 years later.

Visual tour

KEY

▼ **MATTERS OF STATE** Around the map's borders, Lord Burghley has added copious handwritten assessments of the political and administrative problems associated with the Scottish borders, or "Marches". At top left, he lists the "Names of ye principal lordships in the Middle March" who were loyal to Queen Elizabeth I, and the number of horses they could muster to defend the Elizabethan state.

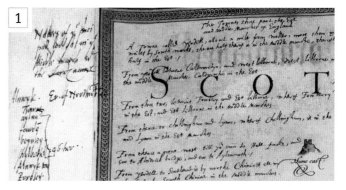

1

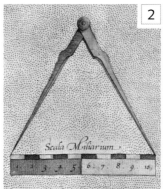

2 ◄ **COMPASS AND SCALE BAR** Saxton depicts a pair of compasses and a scale bar, numbered showing 10 miles (16km) equal to just over 6.5cm (2½in), which suggests an overall scale of around 1:300,000. The absence of Saxton's usual signature from the bar suggests this was an early, unfinished draft of the map sent to Burghley for approval.

3

▲ **DEFENCE AND SURVEILLANCE** Dense with towns, villages, and hills, the map's detailed topography enabled Burghley to see how best to defend the north of England. He would have also used it to keep an eye on troublesome northern earls, such as Henry Percy, 8th Earl of Northumberland, a Catholic sympathizer and supporter of Mary, Queen of Scots, the main challenger to Elizabeth's rule. Percy's seat at Alnwick (centre) is particularly prominent.

Vatican Gallery of Maps

c.1580 ▪ FRESCO ▪ c.3M × 4M (c.10FT × 13FT) PER MAP ▪ VATICAN PALACE, ROME, ITALY

EGNAZIO DANTI

Decorating palaces, monasteries, and villas with cycles of maps first became fashionable in 16th-century Italy, as popes and princes sought ways to display their wealth and learning. The greatest cycle of all is the Gallery of Maps in Rome's Vatican Palace. Designed by a team that included the Dominican friar and mathematician Egnazio Danti, they were commissioned by Pope Gregory XIII to decorate a 120-m (394-ft) corridor in the Belvedere

Courtyard. The cycle includes 40 huge, vividly coloured fresco maps, each showing an Italian region and its islands. Each map measured around 3 × 4m (10 × 13ft), and was positioned to create the illusion of walking through Italy from north to south. The regions of the Tyrrhenian Sea are on one side (to the east) and those of the Adriatic are to the other (to the west), while biblical scenes adorn the curved ceiling.

EGNAZIO **DANTI**

1536-1586

Born Pellegrino Danti in Perugia, central Italy, into a family of painters, Danti studied theology as a young man. He subsequently joined the Dominican Order (a Roman Catholic religious order) in 1555 and changed his name to Egnazio.

Drawn increasingly to the study of mathematics and geography, Danti moved to Florence in 1563 at the invitation of Duke Cosmio I de' Medici. There he taught mathematics and was also commissioned to work on his first great cartographic project – 53 maps, globes, and paintings for the *Guardaroba* (or "wardrobe") in the Palazzo Vecchio, the Town Hall of Florence. Following his success in Florence, Danti was appointed professor of mathematics at the University of Bologna. He was then invited to Rome by Pope Gregory XIII to work on the reform of the Christian calendar – which become known as the Gregorian Calendar. Here, Danti was also given the task of overseeing the designs for a series of regional maps to decorate the newly constructed Gallery of Maps in the Vatican. In addition to his cartographic and mathematical endeavours, Danti was also a greatly respected bishop, and was widely admired for his charitable work with the poor in southern Italy.

> The gallery of maps was a largely unsurpassed synthesis of modern geography and a papal interpretation of church history

FRANCESCA FIORANI, *THE MARVEL OF MAPS*

Visual tour

6 2 3 1

VATICAN GALLERY OF MAPS: 120M (393FT)

4 5

KEY

W
S — N
E

▶ **MAP OF "FLAMINIA"**
The Via Flaminia was a road that stretched from ancient Rome across the Appenine Mountains to Rimini. It is where Julius Caesar famously crossed the Rubicon river in 49 BCE, starting a civil war in republican Rome. This decisive moment in Roman history is represented by Caesar's army, shown crossing the river at the map's centre, and also commemorated on the obelisk.

1

2

▲ **THE BATTLE OF LEPANTO**
At the far end of the gallery, a historical map shows the naval victory of a "Holy League" of Christian forces over the Turkish fleet near Lepanto in the Ionian Sea in October 1571. The victory was widely seen as the result of divine intervention, and the papacy took credit, commemorating it in a new feast day, "Our Lady of Victory".

▶ **MILAN IN HISTORY** Danti's maps often showed important historical events from different eras taking place on the same territory. The map of Milan paid little attention to the city, and instead concentrated on at least three historically important battles that took place in the region: the battle between Hannibal and Scipio in 218 BCE; Charlemagne's defeat of the Lombards in 774 CE; and the French defeat at Pavia in 1525. Here, geographical detail is blended with past events to depict how they shaped regional identities.

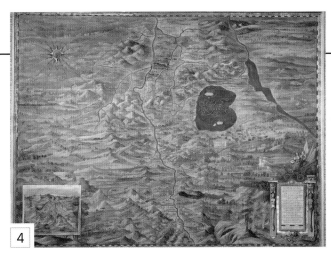

4

▲ **PERUGIA** In the late 1570s, Danti surveyed his homeland of Perugia. The region is shown here in exquisite detail, with cartouches, compass roses, and inset town maps. The map also depicts Hannibal's famous victory over the Romans in 217 BCE at Lake Trasimeno.

5

▲ **CAMPANIA** The fresco of the Campania region has an inset map of Naples. The map shows the Battle of Garigliano (915 CE), when Pope John X's Christian forces defeated a Fatimid Saracen army that had established a colony in the region.

▼ **MAP OF ITALY**
The map cycle begins with two maps of the peninsula either side of the gallery's southern entrance. One shows ancient (*antiqua*), the other modern (*nova*) Italy. They represent the mix of the classical and the contemporary Italian worlds – a trend that defines the rest of the cycle, with its mix of paganism and Catholicism.

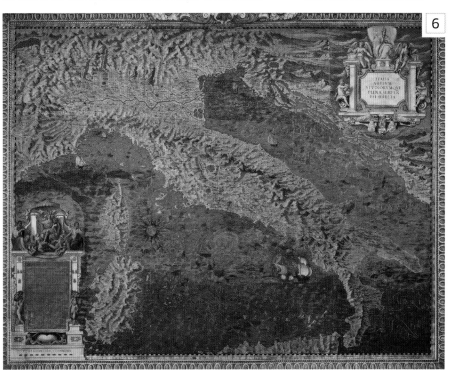

6

IN **CONTEXT**

Danti's maps were produced during a time of great religious turmoil. They were part of a wider Catholic response to Protestantism, as decided by the Council of Trent (1545–63), a meeting convened by the papacy to reform Catholic theology and condemn Protestant "heresies". Some Protestant cartographers had produced popular printed maps of the Holy Land to be used as visual aids in reading vernacular Bibles. So, as part of Pope Gregory XIII's attempt to create a "sacred cartography", Danti's maps offered an image of a universal Catholic Church centred on Italy, Rome, and the papacy (rather than Jerusalem), which was not afraid to use science to extend its influence across the globe.

▶ **The effects of the Council of Trent** permeated into 16th-century politics and culture. This print depicts the council in session.

The Molucca Islands

1594 ▪ COPPERPLATE ENGRAVING ▪ 36.7CM × 53CM
(1FT 2½IN × 1FT 8¾IN) ▪ MITCHELL LIBRARY, SYDNEY, AUSTRALIA

SCALE

PETRUS PLANCIUS

This map represents a quiet revolution in navigational methods and in mapmaking's relations with trade. Its story began in 1592, when the States General, which ruled the Dutch Republic, commissioned astronomer and cartographer Petrus Plancius to produce maps supporting Dutch attempts to enter the southeast Asian spice trade – dominated at the time by the Portuguese. Plancius obtained maps from Portuguese pilots through various nefarious means, employing them to create a series of maps using Mercator's projection (see pp.110–13). Although the projection had been around for over 20 years, its complex mathematics was confusing for navigators reluctant to give up their old portolan charts; for the first time, Plancius introduced it into maps used in navigation. This map of the prized spice-producing islands of the Moluccas, correcting their misplacement on Ribeiro's map (see pp.100–103), was particularly important in showing a new generation of Dutch merchants the spices and other riches lined that awaited them if they invested in such new navigational methods.

In 1595, the first Dutch voyage to Indonesia set off. Prior to its departure, Plancius acted as its scientific consultant, preparing maps such as this to enable the fleet to navigate its way between the myriad islands. The map represents a massive leap in understanding, both of the Spanish-controlled Philippines to the north, and of its geographical and commercial centre of gravity, the Moluccas, shown running along the Equator. New Guinea is sketched in, although without its undiscovered southern coast, as well as a mysterious place labelled "Beach". The map is a unique mix of sailing chart and commercial advertisement, and anticipates the rise of the Dutch East India Company (often known as the VOC, after its Dutch name, *Vereenigde Oost-Indische Compagnie*).

PETRUS **PLANCIUS**

1552–1622

Born Pieter Platevoet, Dutch astronomer and cartographer Petrus Plancius produced over 100 maps for the Dutch East India Company.

A theology student, Plancius fled religious persecution in the Spanish-controlled southern Netherlands for Amsterdam, where he became a minister in the Dutch Reformed Church and developed his mapmaking skills. He made pioneering astronomical observations, named new constellations, proposed innovative methods for determining longitude, and made celestial and terrestrial globes. His mapmaking experience made him an ideal scientific adviser to the state-sponsored attempts to expand Dutch overseas trade, which came to dominate southeast Asia.

The first printed Dutch charts showing seas and coastlines outside Europe

GÜNTER SCHILDER, DUTCH CARTOGRAPHIC HISTORIAN

Visual tour

KEY

▼ **JAVA** In the 1590s, when the Dutch were trying to break into the spice trade in the Indonesian archipelago, accurate maps of key trading islands such as Java were essential. Plancius includes shoals and rocky areas as a guide for inexperienced pilots sailing in the area for the first time.

▼ **SEA MONSTERS** For all its confident scientific techniques, the map still depicts sea monsters roaming the Pacific Ocean. This one looks like an outsized fish, showing more of an interest in decoration than a belief in aquatic monstrosity.

▲ **NUTMEG** Across the bottom of the map, Plancius has drawn the precious commodities sought by the Dutch merchants and pilots. These images of "nux myrstica" (nutmeg), together with similar ones of sandalwood and cloves, allow users of the map to identify both plants and fruit with ease. Ownership of the map promises access to such riches.

▶ **NOVA GUINEA** The island of New Guinea dominates the map, but Plancius has confused its western coast and outlying islands by separating the peninsula of Irian Jaya (or West Papua) from the mainland and joining it to the island off to the left. This shows how cartographic errors can develop from the circulation of hazy geographical information.

5

INSVLAE MOLVCCAE celeberrimæ
sunt ob Maximam aromatum copiam quam per totum terrarum orbem mittunt: harum præcipuè sunt Ternate, Ti-
doris, Motir Machian et Bachian, his quidam adjungunt
Gilolum, Celebiam, Borneonem, Amboinum et Bandam,
Ex Insula Timore in Europam advehuntur Santala rubea
& alba, Ex Banda Nuces myristicæ, cum Flore, vulgò dicto,
Macis, Et ex Moluccis Caryophilli: quorum icones in
pede hujus tabellæ ad vivum expressas poni curavimus .

Hispanicæ leucæ 17½ uni gradui competentia.

Miliaria Italica 70 singulis gradibus respondentia

Miliaria Germanica, quorum 15 uni gradu respondent

Visscher excudebat A° 1617.

Baixos de S.
Bartholome

I: de S: Petro

◄ **CARTOUCHE** The beautifully engraved cartouche describes the region and the commodities that travellers could obtain throughout the islands. The scale bar beneath enables accurate navigation and suggests the union between scientific measurement and new ways of doing business. The cartouche's elaborate form derives from designs for funerary monuments.

▼ **THE MOLUCCAS** Along the Equator lie the Moluccas. To the left is Sulawesi (labelled "Celebes"), in the middle Halmahera ("Gilolo"), to the right Seram ("Ceiram"), and below lies Ambon. Plancius placed them all too close together, as if he wanted to unite them in greater commercial harmony.

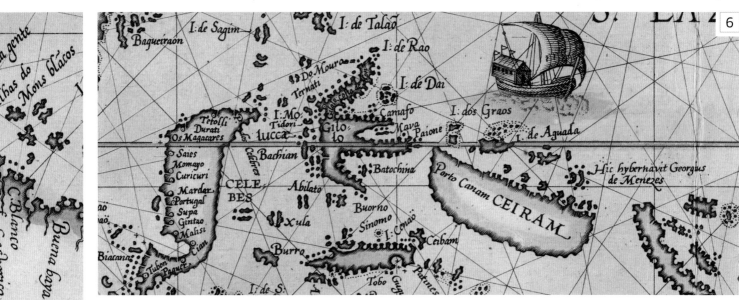

6

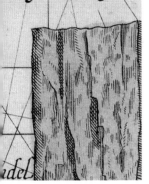

7

▲ **THE MYSTERY OF "BEACH"** On the map's southernmost edge is the mysterious "Beach", a name that is a garbled version of Marco Polo's Siam, the Chinese name of which had been mistranscribed as "Boeach". This land mass could either reflect older medieval geography or be a glimpse of Australia, decades before its official discovery.

ON **TECHNIQUE**

Most sailing charts made before the 17th century treated the Earth as though it were flat. This meant that they could not correctly show direction or distance over large areas. Plancius's use of Mercator's projection enabled more accurate sailing. The lines of latitude are spaced further apart as they move away from the Equator, enabling Dutch pilots to plot accurate straight lines across the map. This made sailing easier and safer, and enabled merchants to trade more reliably, and hence more profitably, in areas thousands of miles from home.

▲ **This 17th-century** Dutch East India Company map shows navigational routes around the Straits of Magellan.

Map of the Ten Thousand Countries of the Earth

1602 ■ WOODBLOCK PRINT ON MULBERRY PAPER ■ 1.82M × 3.65M (5FT 11½IN × 12FT)
■ JAMES FORD BELL LIBRARY, MINNEAPOLIS, MINNESOTA, USA

SCALE

MATTEO RICCI, LI ZHIZAO, AND ZHANG WENTAO

Known as the "**Impossible Black Tulip**" of cartography because of its rarity, beauty, and exoticism, Matteo Ricci's map is a remarkable fusion of early 17th-century European and Chinese geographical knowledge, which offered China its first glimpse of America. Ricci, an Italian Jesuit leading the Society of Jesus's missionary work in

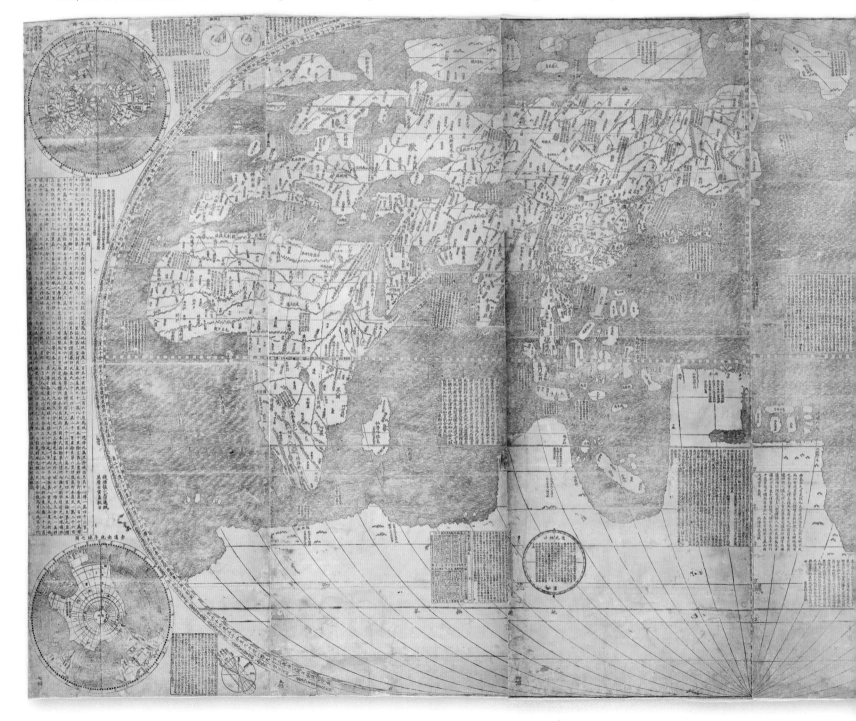

China, was also a gifted scientist who believed that understanding Chinese culture and learning was central to his goal of converting the locals to Christianity.

A Chinese-European collaboration

In 1584, Ricci established a Jesuit mission in Zhaoqing, southern China. He drew a map on the wall using European characters, and when admiring Chinese officials requested that he made the map "speak Chinese", he drew a new version using Mandarin. Both maps have since been lost, but they provided the prototype for this one, begun in 1601 when Ricci travelled to Beijing at the invitation of the Ming emperor Wanli. It was a truly collaborative project, published by the Ming printer Zhang Wentao on six huge wood blocks using brown ink on mulberry paper,

a method used for making large screens to be displayed in semi-public places. Ricci was also assisted by the renowned Ming mathematician and geographer Li Zhizao who, following the map's completion, became one of Ricci's most celebrated converts to Christianity.

Ricci's oval projection is taken from the Flemish cartographer Abraham Ortelius's famous 1570 world map in his atlas *Theatre of the World*. He also borrows from other European mapmakers, including Mercator (see pp.110–13) and Plancius (see pp.122–25). These are fused with Chinese sources provided by Li Zhizao, which gave Ricci the closest insight into the country's geography ever afforded a European. Strikingly innovative, it places China near the centre of the map, and introduces a graticule (coordinate grid), which was unfamiliar to Chinese eyes. The Chinese admired it for its novelty and use of astronomical and cosmographical observations, which were so important for Ming policy; Ricci used it to convince the Chinese of the primacy of a Christian God capable of creating such a world.

MATTEO **RICCI**

1552-1610

Known as the "Apostle of China", Ricci was trained in Rome as a Jesuit. He entered the order in 1571 and began his missionary work in the Portuguese colony of Goa, India, in 1578, before moving to China in 1582.

Ricci started work in Macau, at that time the centre of Christian missionary efforts in China. He learned the Chinese language and began expanding his religious mission. Ricci then travelled to Zhaoqing in Guangdong Province, southern China, where he began mapmaking and compiling the first dictionary transcribing Chinese into a European language (Portuguese). Appointed Major Superior of the Jesuit mission in China, Ricci was subsequently invited to Beijing in 1601 and appointed adviser to the Ming emperor Wanli. He established the city's oldest Catholic church, and worked on his world map alongside astronomical projects.

Visual tour

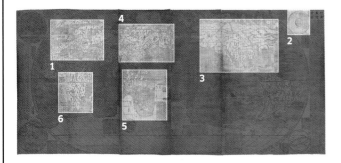

KEY

> ▶ **EUROPE** Although it is described here for the Chinese reader in glowing terms – pious, rich, and powerful – Europe is drawn surprisingly badly. The Rhine flows from the Danube and an unidentified river connects the Black Sea with the Baltic. Ricci's notes are little better: according to him, dwarves live in the northeast, while St Patrick's legendary banishment of snakes from Ireland is ascribed to England.

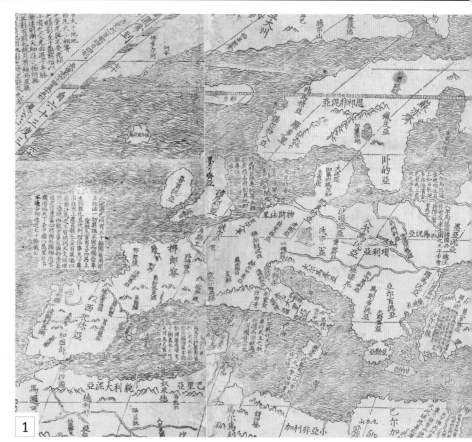

1

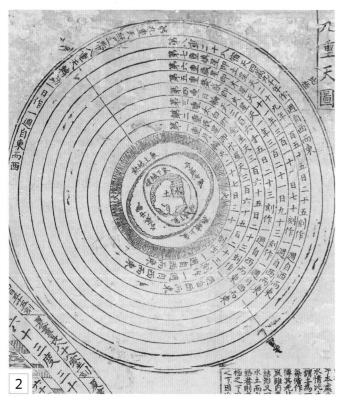

2

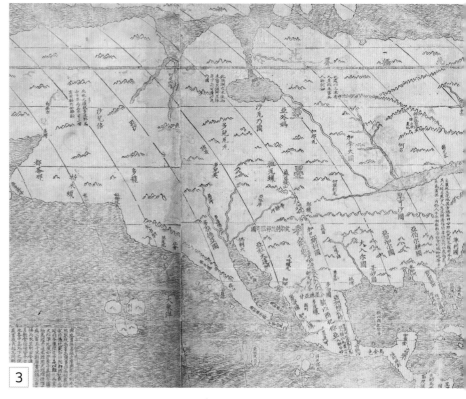

3

▲ **DIAGRAM OF THE PLANETARY SYSTEM** In line with Jesuit teaching, Ricci's cosmological diagram ignores Copernicus's heliocentric theories. Instead, Ricci's resolutely geocentric system, based on Ptolemy, locates the Earth at the centre of the universe – happily coinciding with Chinese beliefs. This diagram even shows China in the middle of the tiny terrestrial globe. Its nine concentric circles represent the planets, culminating in the fixed heaven. The Earth is surrounded by two oval rings of air from cold and warm regions, enveloped by fire.

▲ **NORTH AMERICA** America is shown to a Chinese audience for the first time, although its accuracy is rather questionable. A lake incorrectly stretches from the Arctic to the St Lawrence River; although Labrador is named correctly. Florida is described as the "Land of Flowers", while in another region called "Ka-na-t'o-erh" (possibly meaning Canada), "the inhabitants are excellent" and "kind to strangers", but their mountainous neighbours "kill, fight, and rob one another all year round".

▼ **CHINA AND JAPAN** Ming China regarded itself as *Zhōngguó*, (The Middle Kingdom), the political and cultural centre of the world and is shown here in great detail along with Korea and Japan. Ricci adjusted Ortelius's prime meridian so it ran through the Pacific because he believed the Chinese objected to "our geographies pushing China into one corner of the Orient".

▼ **MALACCA AND THE SOUTHERN CONTINENT** Ricci's grasp of his Chinese sources often seems confused: Malacca, he says, "abounds in flying dragons which coil around trees". The Malay Peninsula contains the first description of a cassowary, a rare bird, while the southern continent is named "parrot country" after Plancius (see pp.122–25), and alludes to European discoveries on the Australian coast.

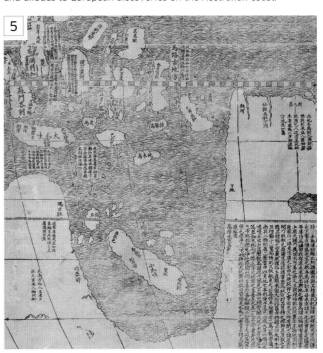

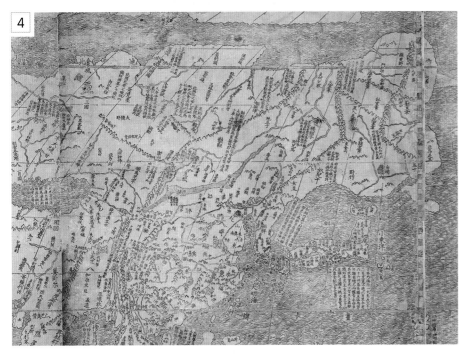

◀ **SOUTHERN AFRICA** Following Portuguese voyages round the Cape of Good Hope (a voyage Ricci experienced himself in 1578), the map shows southern Africa in reasonable detail. It gives southeast Africa its Portuguese name of "Monomotapa" and Ricci cannot resist speculating on the existence there of "an animal with a head like a horse, a horn on his forehead", concluding, "One wonders if it is a unicorn".

IN **CONTEXT**

Almost as soon as the Jesuit's Society of Jesus was founded in 1540, China was identified as a place for missionary work. One of the Society's founders, St Francis Xavier, died trying to establish the first Chinese mission in 1552, before its formal establishment in Macau in 1563. Ricci and his fellow Jesuits attempted to pursue a policy of conversion that involved complete assimilation within Chinese culture, from learning its language to adopting its dress codes. The result was a remarkable period of east-west cultural and scientific change, led by the brilliant Ricci.

▲ **Matteo Ricci** pictured with his first convert in China.

The Selden Map

c.1608–1609 ▪ INK ON PAPER ▪ 1.58M × 96CM (5FT 2¼IN × 3FT 1¼IN) ▪ BODLEIAN LIBRARY, OXFORD, UK

UNKNOWN

SCALE

Old maps are constantly being discovered – many, as in the case of this beautiful, enigmatic map whose secrets are only now being revealed, found languishing in dusty libraries. *The Selden Map* was rediscovered at the beginning of the 21st century in the basement of Oxford University's Bodleian Library, where it had been sitting neglected for nearly a century. It is now regarded as the most significant Chinese map of the last 700 years.

The map shows the whole of southeast Asia and its maritime sea routes at a scale and in a style unknown in any comparable Asian map of the period. Virtually nothing is known about its makers, although it is thought to have been made around 1608–09 in the late Ming dynasty. The map first entered the Bodleian Library in 1654, when it was bequeathed as part of the collection of the English scholar, John Selden, which is how it acquired its name.

Oceanic trade in the Ming dynasty

Exquisitely drawn and painted with black ink on enormous sheets of paper – glued together to make one vast piece, possibly to hang on a wall – the map is oriented in usual Chinese style with north at the top. It shows southeast Asia centred on Ming China, stretching from the Indian Ocean in the west to the Spice Islands in the east, including the Moluccas (see pp.122–25). To the north is Japan, and to the south, Java. Beyond these obvious elements, the map poses a series of puzzles. It has a European-style compass rose and scale bar, which were unknown on Chinese maps for centuries. Instead of putting Ming China in the middle – the Chinese word for China, *Zhōngguó*, means the "middle kingdom" – the chart's creators have made the unprecedented

decision of placing the South China Sea at its centre. The map's main emphasis is on a series of seaborne trade routes plotted using compass bearings, shown radiating outwards from the port of Quanzhou, on the eastern Chinese coast (near Taiwan, in the middle of the map). These shipping routes reach as far as Calicut in India, shown on the map's western edge, and also describe how to sail to commercially important locations such as Oman on the Arabian Peninsula and Hormuz in the Persian Gulf. This appears to be a chart centred on the sea, not the land and, as such, it represents the beginning of a whole new era of cartography and of the use of maps in Ming-dynasty China.

JOHN **SELDEN**

1584–1654

Selden was an English scholar specializing in legal history, as well as a historian, an antiquarian, and a politician.

It is not known how or when Selden acquired this map, but he was a renowned polymath. In particular, he was a distinguished law scholar and a prodigious collector of manuscripts, especially those with oriental origins. Selden was one of the first English scholars to show an interest in Persian, Arabic, and Chinese learning, and his work inspired others. When he died, Selden bequeathed more than 8,000 manuscripts to the Bodleian Library at his alma mater, Oxford University. The collection includes not only this map, but also the *Codex Mendoza* (see pp.104–105).

A rich cartographical image of the economic dynamism of 17th-century East Asia

ROBERT BATCHELOR, *THE SELDEN MAP REDISCOVERED*

Visual tour

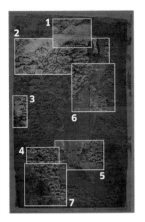

KEY

◀ **COMPASS AND SCALE BAR** 1
Among the map's many unique features are its compass rose and scale bar – elements that are absent from earlier Chinese maps. The compass rose has 72 points, and as if to emphasise its novelty, *luojing* (compass), is written in the middle. The scale bar is also unprecedented. It represents one Chinese *fen* (foot) at a scale of approximately 1:4,750,000. Both of these features demonstrate that the Chinese cartographers possessed extensive knowledge of European maps.

▶ **BEIJING AND THE GREAT WALL** In keeping with Chinese tradition, the city of Beijing is prominently marked on the map, as is the Yellow River (below left) and the provincial boundaries (which look more like river channels). Across the centre run the crenellated fortifications of the Great Wall of China, stretching from Shanghaiguan to Lop Lake. Much of the existing wall shown here was constructed by the Ming Dynasty and was begun as early as the 7th century BCE.

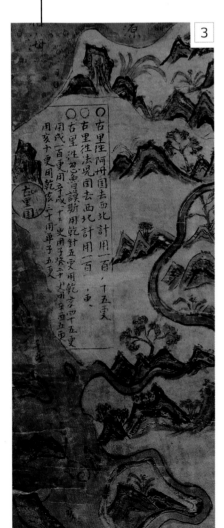

3 ◀ **CALICUT AND THE INDIAN OCEAN** On the far left lies Calicut, a port on the west coast of India, and one of the most important trading centres in the Indian Ocean. The Bay of Bengal is absent, but notes describe how to sail from Calicut to Yemen, Oman, and Hormuz. This suggests the map's maker was drawing on descriptions of the early 15th-century voyages across the Indian Ocean by Ming-dynasty explorer Zheng He (see pp.134–37).

4 ◀ **SUMATRA**
The island of Sumatra, in modern-day western Indonesia, is depicted as another key location. It was the westernmost point in the network of trading routes across the South China Sea, the entrance into the Indian Ocean, and a pivotal location in the fierce rivalry between the Dutch, English, and Portuguese over control of the spice trade.

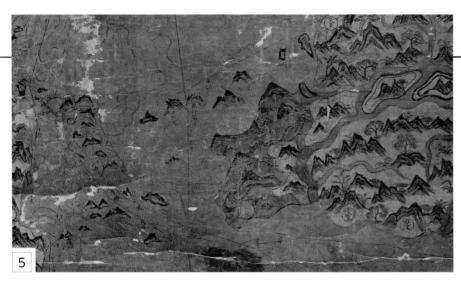

5

▲ **THE MOLUCCAS** Situated at the confluence of a series of maritime trade routes running north, south, and west, sit the spice-producing islands of the Moluccas (see pp.122–25). These lush, tropical volcanic islands, shown here in the eastern part of modern-day Indonesia, were central to the economy of the entire region.

6

▲ **TRADING PLACES** The east coast of China, centred here on Quanzhou, is the map's commercial centre of gravity. From here, a spidery network of trade routes can be seen heading off to all four points on the globe, connecting the Ming empire to the rest of the world's commerce.

▶ **ARTISTRY** Even relatively peripheral details, such as the islands at the westernmost point on the map, are rendered with loving detail. As much as this is a map about maritime trade and global commerce, the unknown mapmaker also pays close attention to the flora and fauna of the region.

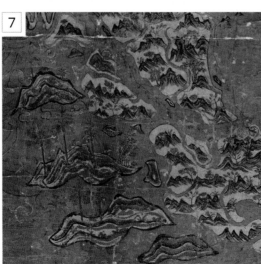

7

IN **CONTEXT**

Chinese navigators had been using *luojing* (compasses) that pointed *zhinan* (south) since at least the 10th century. Some were "dry-pivots", attached to a post, whereas others were floating compasses, where a magnetized needle was placed in a basin of water. Readings were used to draw up *zhenjing* (compass manuals), the equivalent of European "rutters" – written descriptions of how to sail from one place to another, listing ports, islands, and currents, all based on compass readings. Most Chinese maps, like Selden's, have north at the top because imperial subjects faced upwards to the emperor, who looked "down" (south).

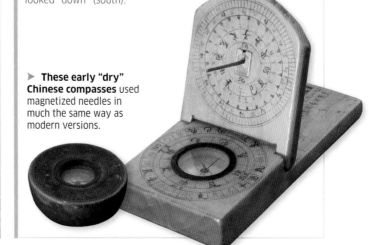

▶ **These early "dry" Chinese compasses** used magnetized needles in much the same way as modern versions.

Nautical Chart

1628 ▪ WOODBLOCK PRINT ON PAPER ▪ 10CM × 14.5CM (4IN × 5¾IN) PER PANEL
▪ LIBRARY OF CONGRESS, WASHINGTON, DC, USA

SCALE

ZHENG HE

Between 1405 and 1433, the Chinese Ming dynasty dispatched Admiral
Zheng He on one of history's most ambitious series of maritime expeditions,
stretching 12,000km (7,500 miles) from Nanjing in eastern China to Hormuz
in the Persian Gulf and Mombasa on the east African coast. Zheng He's seven
voyages were vast logistical operations that involved hundreds of ships
and thousands of soldiers, motivated by various imperial, commercial, and
diplomatic ambitions. Although maps and charts from the original voyages
have not survived, a later version was reproduced in a vast compendium
of military technology and preparations, entitled *Wubei Zhi*, written by the
Ming officer Mao Yuanyi (c.1594–c.1641).

Exploring the oceans

The charts draw mainly on Zheng He's last voyage in 1431–33, but also
include data accumulated over the previous expeditions. Originally designed
as a strip map measuring 20cm × 5.6m (7¾in × 18ft 4¼in), they were cut down
to fit 40 pages in the *Wubei Zhi*. Running from right to left, the charts begin
in Nanjing and end in Africa, naming 530 places as they move east to west,
although the orientation and scale shift constantly according to the importance
and information gleaned from specific regions. The coverage of particular
regions also reflects Chinese preoccupations: the Ming territories are given
18 pages and southeast Asia has 15, while Arabia and East Africa have just six.
More than 50 sea routes cover the maps and are shown as dotted lines, often
supplemented by detailed sailing instructions based on compass bearings,
providing a wealth of information on ports, coastlines, and islands, as well as
the depth and flow of water. Despite the charts' many distortions due to their
size, they are a remarkable record of one of the great and often overlooked
periods of seaborne exploration.

ZHENG **HE**

c.1371–1433

Born into a Muslim Hui family in Yunnan Province, Zheng He was
castrated by an invading Ming army in the 1380s, before entering service
in the royal household.

Zheng He served as a eunuch to Zhu Di, the future Yongle Emperor. He led
several of Zhu Di's military campaigns against internal and external forces,
and, in 1405, was appointed admiral of the first of seven seaborne voyages.
These were intended to impose the Ming dynasty's commercial control over
the Indian Ocean, and develop relations with Arabia and Africa. In July 1405,
a fleet of more than 300 ships and 28,000 crew left China, travelling between
southeast Asia, India, the Arabian Peninsula, and eastern Africa. Zheng He's
expeditions dwarfed those of later Portuguese and Spanish voyages.

> We have… beheld in the ocean, huge waves like mountains rising sky-high, and we have set eyes on barbarian regions far away

ZHENG HE

Visual tour

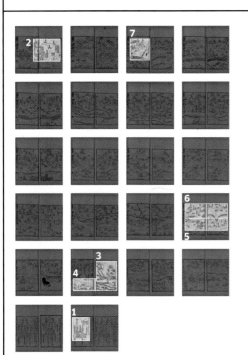

KEY

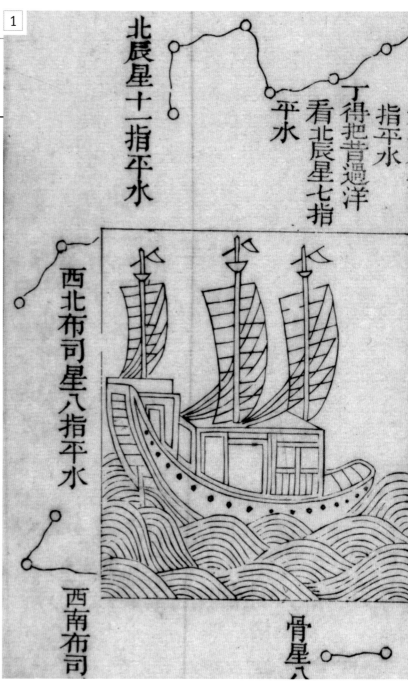

▶ **NAVIGATING BY THE STARS** The book culminates in an exquisitely drawn three-masted vessel, one of the smaller ships used on Zheng He's expeditions. It is surrounded by commentaries of just five of the many constellations that were used by Zheng He's pilots to navigate across the ocean. The map records that their voyage has been successful thanks to accurate astronomical observations.

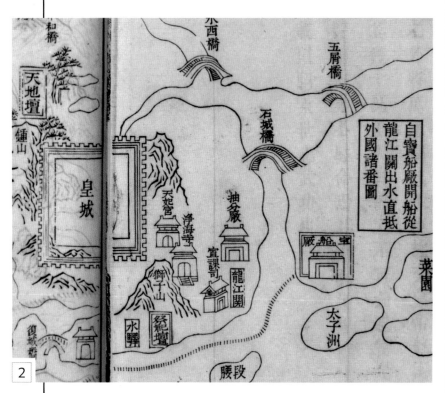

▲ **NANJING** In 1368, the first Ming emperor Hongwu founded the dynasty's capital at Nanjing on the Yangtze River. This is represented by the square cartouche on the left, and several identifiable imperial buildings. Across the river by the imperial shipyard, a dotted line marks the departure of Zheng He's fleet in 1405.

▶ **INDIA, AFRICA, AND SRI LANKA** This section, oriented with north to the left, has southern India, including Cochin, at the top, Sri Lanka to the right, and the Maldives to the bottom left. In 1409, Zheng He erected a trilingual stone tablet in Chinese, Tamil, and Persian in Galle, Sri Lanka, listing his dedications to Islamic, Hindu, and Tamil deities.

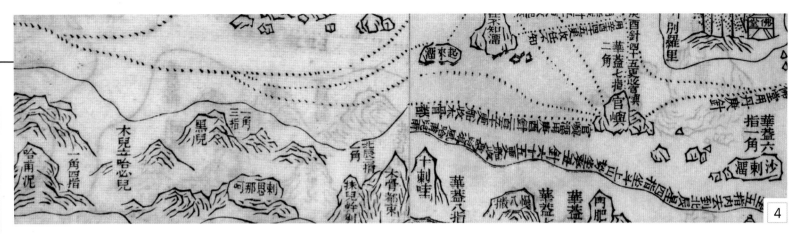

THE CHINESE IN EASTERN AFRICA The sea routes and ports along the eastern African coast run from right to left (or south to north), with the Maldives shown at the top. The map is dominated by Mozambique to the bottom right and Mombasa to the left. Zheng He first visited the region on his fifth voyage in 1418.

SUMATRA Western Sumatra is drawn here at the confluence of a series of sea routes. It was a pivotal commercial location between South China and the Indian Ocean, and is where Zheng He stored trade and fleet goods in 1413.

IN **CONTEXT**

Chinese maritime navigation drew on a long and distinguished astronomical tradition, which enabled admirals like Zheng He to use reliable stellar observations to navigate across the Indian Ocean. The colossal size of his fleet allowed him to travel with teams of experts who could assist with practical and astronomical navigation. It is claimed the largest of the so-called "Treasure Ships" that carried Zheng He and his deputies were over 100m (330ft) long, although around 60m (200ft) is now considered a more realistic estimate. With nine masts and four decks, these were some of the largest wooden ships ever built. The fleet also included warships, patrol boats, and troop and horse transports in one of maritime history's most logistically ambitious expeditions.

BURMA Zheng He's charts show coastlines in great detail, including those of Burma, which would have been familiar territory to a Ming commander. The region of Tenasserim lies to the top right, with pagodas to the left, while the Nicobar Islands stand off the coast, with the compass route to India leading off to the left.

THE EAST CHINA SEA
The chart's scale and orientation shifts constantly. The East China Sea around Shanghai and the Huangpu River is on a very large scale, and shows ports, flag poles, and coastal features. Precise terms describe sailing routes off the coast, with west at the top.

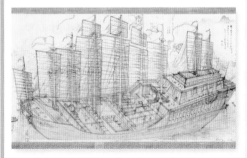

This illustration depicts a magnificent 15th-century Chinese ship, similar to the kind that Zheng He would have used.

Map of the "Inhabited Quarter"

1647 ▪ INK ON PAPER ▪ 14CM × 26CM (5½IN × 10¼IN) ▪ BRITISH LIBRARY, LONDON, UK

SCALE

SADIQ ISFAHANI

Just one of 33 maps in an atlas called the *Inhabited Quarter*, the Indian scholar Sadiq Isfahani's hand-drawn world map is part of an even larger encyclopaedic work, the *Shahid-i Sadiq*, dated 1647. A fusion of Greek, Indian, and Islamic geography, it centres on the "Inhabited Quarter", the northern half of the eastern hemisphere which, apart from its southern orientation, bears a striking resemblance to Ptolemy's inhabited world (the *ecumene*: see p.24).

The world sits in a bowl. Europe is at the bottom right, Africa top right. The Indian Ocean runs across the top, containing two large islands, presumably Sri Lanka and Madagascar. India and China are named but drawn very crudely. Perhaps the most striking feature is the map's graticule, a grid of equal squares representing parallels and meridians, defined primarily by the Greco-Islamic tradition of seven *klimata* or climes (see p.43) running from east to west.

SADIQ **ISFAHANI**

c.1607–1650

Sadiq Isfahani was born in Muslim Mughal India, probably in Jaunpur in the modern state of Uttar Pradesh, a major centre of Urdu and Sufi scholarship.

Little is known about Sadiq Isfahani's life beyond his writings. His father held important administrative positions at the imperial court of the Mugal emperor Jahangir (reigned 1605–27) which he seems to have bequeathed to Sadiq Isfahani. He, in turn, worked for Jahangir's son and successor, Shah Jahan (reigned 1628–58), who oversaw a golden Age of Mughal art, architecture, and science. Isfahani spent most of his career working in Bengal. His writings took the form of huge, encyclopaedic works composed in Persian. They included a four-volume general history of Asia, entitled *Subh-i-Sadiq* (1646), and the *Shahid-i Sadiq* (1647), which features the atlas and map shown here.

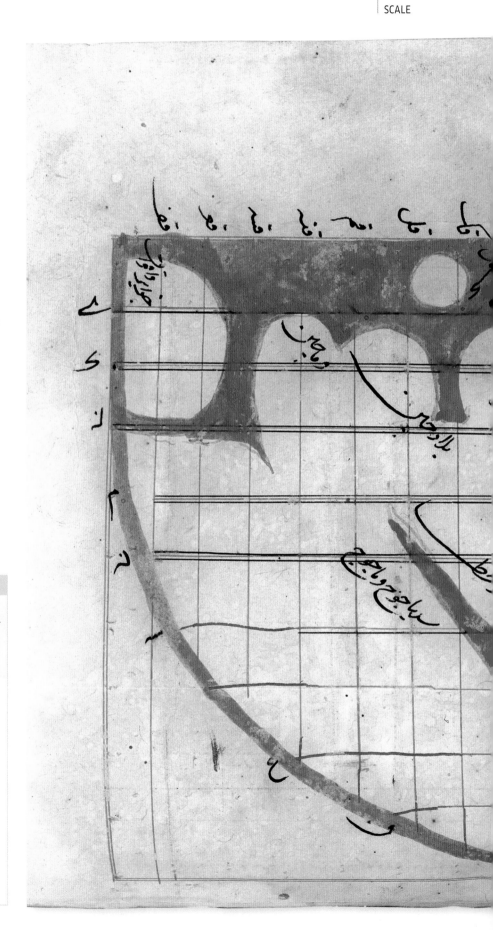

An attempt to present the **political and economic geography of the Indian subcontinent** during the heyday of the Mughal empire

TAPAN RAYCHAUDHURI, INDIAN HISTORIAN

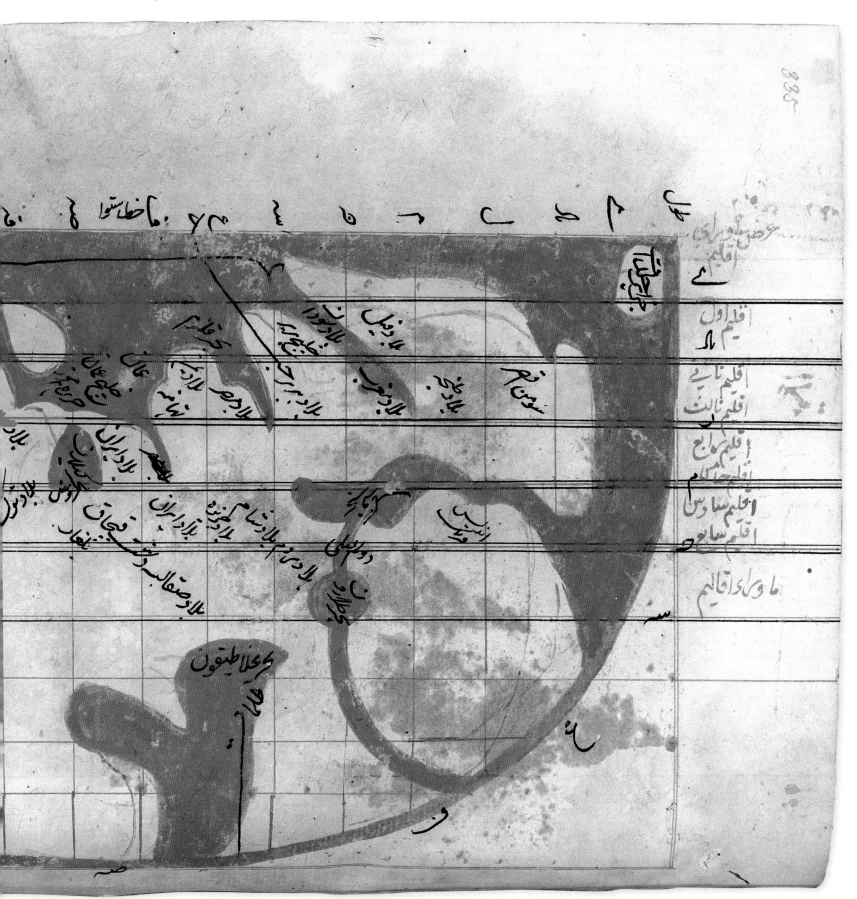

Visual tour

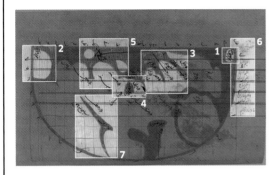

KEY

1

▶ **THE CANARY ISLANDS**
In keeping with earlier western geographical traditions going right back to Ptolemy, the map measures longitude starting from the Canary Islands, situated on the map's furthest edge. However, it shows little awareness of the extent of the Atlantic, or the Americas, nearly 200 years after their discovery.

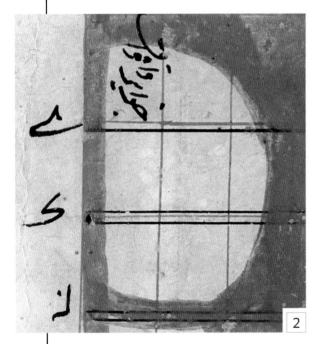

2

▲ **THE MYSTERIOUS WAQWAQ ISLANDS** In the far east lie the Waqwaq Islands, shown in Islamic geography as either a real or a fanciful place, where trees supposedly bore women as fruit. It was believed to be an island or archipelago in east Africa (including Madagascar), or as is more likely here, somewhere in southeast Asia such as Sumatra.

▶ **AFRICA, LAND OF ELEPHANTS** Southern Africa is very schematic, and is labelled the "Land of Elephants". This reproduces the descriptions and illustrations of many earlier Christian and Muslim mapmakers, although its bizarre, horn-like tip pointing eastwards bears little relation to reality. The Arabian Peninsula is similarly distorted, with Yemen labelled just to the left.

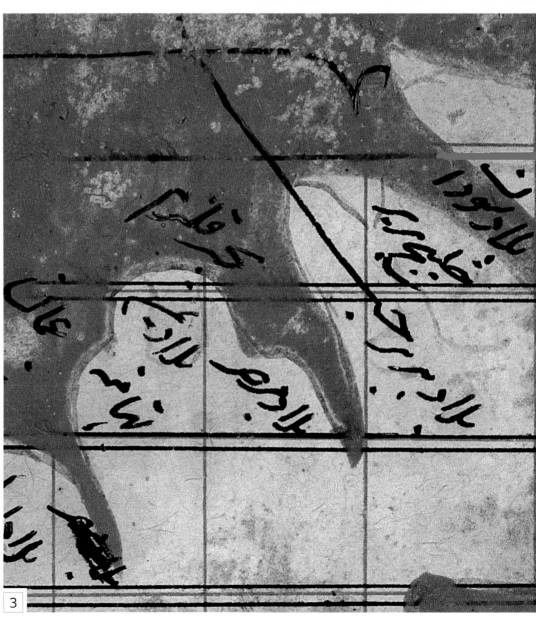

3

4

▲ **THE CASPIAN SEA** Having been made in India, the map understandably has central Asia in more detail than anywhere else. At its centre is the triangular Caspian Sea, where Persian, Greek, Indian, Arabic, and Turkic traditions all meet. Iran is to the right, and below left are the Kipchak Desert (in Uzbekistan), Russia, and Turkestan (today a region of China).

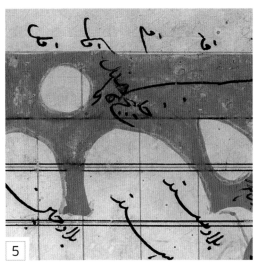

5

▲ **LAND OF SIND** India is clearly labelled, as is "Sind" in modern Pakistan. The region is barely recognizable as a peninsula; Sri Lanka and another island, possibly Madagascar, are shown close to India's west coast.

IN **CONTEXT**

Mughal culture acknowledged a variety of beliefs, and this map uses Ptolemy's method of calculating the seven *klimata* (climes – see p.43) according to the lengths of the longest days in each region. It also adopts Ptolemy's intersecting lines of latitude and longitude, but unlike his curved meridians, Isfahani shows them as straight lines, creating the illusion of a flat Earth. Maps like this aimed to reconcile conflicting geographical traditions that regard the world as divided into habitable or uninhabitable regions, according to climate.

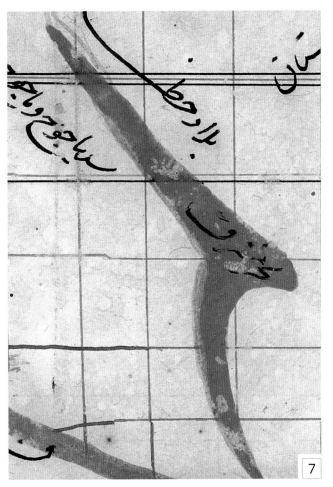

7

◄ **THE SEVEN CLIMES** The map's rim labels the seven climes that divide the world (see box, right, and p.43). As in Greek and Islamic tradition, it shows climes from the first in the torrid far south and the "temperate" climes in India and central Asia, to the cold northern seventh clime.

6

▲ **MORE MONSTERS** Isfahani is not immune to the belief in monstrous races in central Asia. Here he describes Gog and Magog's mythical wall (see p.41, p.43) near a long, hook-shaped sea. This could be Lake Baikal in Siberia, one of the largest freshwater lakes in the world.

▲ **The fifth clime is shown here** in a 15th-century Arabic manuscript known as the *Kitab al-Bulhan*.

New Map of the Whole World

1648 ■ COPPERPLATE ENGRAVING ■ 2.43M × 3M (6FT 8½IN × 9FT 10IN)
■ HARRY RANSOM CENTER, AUSTIN, TEXAS, USA

JOAN BLAEU

SCALE

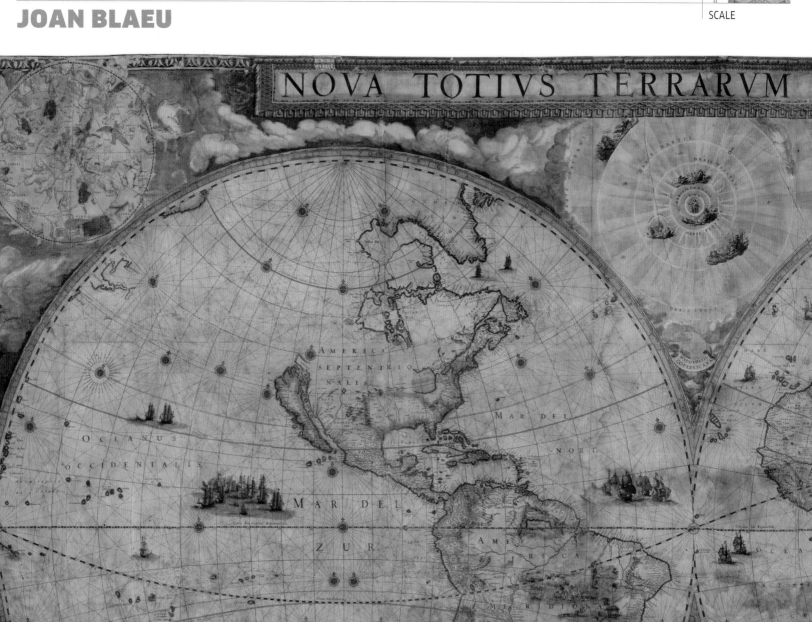

Joan Blaeu's world map is one of the great works of the golden age of Dutch mapmaking, a time of exuberant grandeur that coincided with the Baroque period in the decorative arts. Entitled *Nova totius terrarum orbis tabula* ("New Map of the Whole World"), it features a vast amount of ornamentation that almost overwhelms the geographical detail. It is also a strangely overlooked map: it was the first to reject the centuries-old belief in a geocentric universe with the Earth at its centre, around which circled the Moon, Sun, and other planets. Instead, it endorsed the heliocentric theory of the solar system, proposed by Polish astronomer Nicolaus Copernicus (1473–1543), modified by Danish astronomer Tycho Brahe (1546–1601), and by then widely accepted.

The map was made by Joan Blaeu, mapmaker to the Dutch East India Company (known in Dutch as the *Vereenigde Oost-Indische Compagnie* or VOC), to celebrate the end of the Eighty Years' War (1568–1648) against Spanish rule. It also marks the rise of the Dutch as the new trading superpower of the day. Blaeu used his unique access to the VOC's discoveries to reject classical geography and redraw the globe using contemporary, verifiable information. The map depicts a world coming slowly into view, with incomplete outlines of northwest America and Australasia, and a twin-hemispherical projection that separates the Old World of Europe, Africa, and Asia from the New World of the Americas. As well as its bold statement of heliocentrism, the map is important for recording the first Dutch discoveries in Australia, including Tasmania.

JOAN **BLAEU**

c.1599–1673

Born in the Dutch town of Alkmaar, Joan Blaeu was the scion of a cartographic dynasty that charted the rise of the Netherlands as the foremost trading and exploratory nation of its day.

Joan's father Willem Janszoon Blaeu started a successful printing and mapmaking business in the early 17th century. Although Joan originally trained as a lawyer, he soon abandoned his law career to join the family business. In the 1630s, he created a series of increasingly ambitious atlases and in 1638, following his father's death, Joan took over from Willem as official chartmaker to the Dutch East India Company. He used his position to dominate Dutch mapmaking, and expanded his father's Amsterdam printing works into one of the largest in Europe. Following his 1648 world map, he began work on a vast, 11-volume *Atlas Maior*, which included more than 3,000 pages of text and around 600 illustrated maps. Completed in 1662, it remains one of the largest books ever printed.

Visual tour

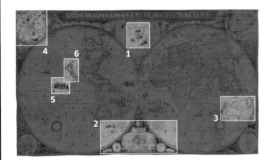

KEY

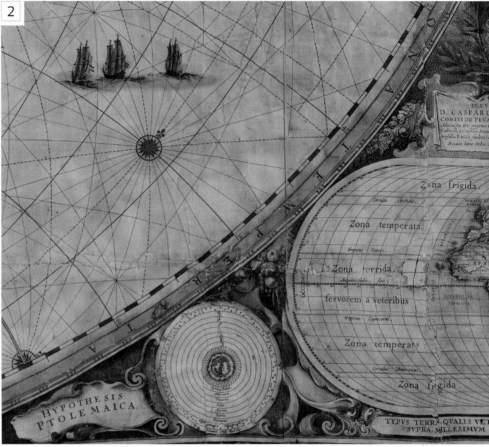

▼ **COMPETING COSMOLOGIES** The small world map just below the centre of Blaeu's projection shows the known world in 1490. The diagram on the left shows the geocentric Ptolemaic cosmos, contrasted with Tycho Brahe's theory of a "geo-heliocentric" cosmos on the right. This remarkable inset map positioned between the two hemispheres captures the changing 17th-century understanding of geography and cosmography.

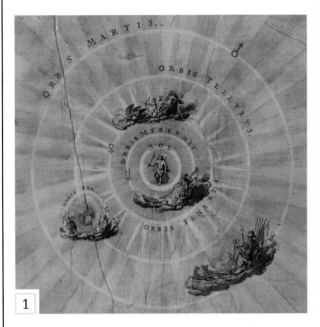

▲ **A HELIOCENTRIC SOLAR SYSTEM** For the very first time in European history, a map follows Copernicus's beliefs in showing the world as part of a heliocentric solar system, with the Sun – personified as a radiant figure named "Sol" – rather than the Earth, at its centre. The outer concentric circles represent the other planets.

▲ **A SOUTHERN CONTINENT EMERGES** The coastline of the western half of mainland Australia is traced for the very first time, as is the island of Tasmania. Australia is labelled "New Holland, discovered in 1644". It remains incomplete, in anticipation of further discoveries along the east coast.

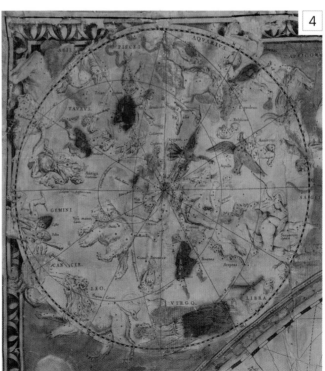

◄ **COSMOGRAPHY – THE BIGGER PICTURE** Renaissance mapmakers preferred to be called cosmographers, in acknowledgment of their apparently magical ability to map the heavens as well as the Earth. Blaeu was no different, although he has relegated his depictions of the cosmos to the top left- and right-hand corners of the map. These show the night sky's constellations as well as the zodiacal signs. Clearly the old beliefs that the stars above influenced the Earth below still held sway.

▶ **GLOBAL TRADE, EAST TO WEST** In the Pacific, north of the Equator, a fleet of European commercial vessels ply their trade, demonstrating just how far global commerce had permeated by the 1640s.

5

IN **CONTEXT**

One of the many claims to fame for Blaeu's 1648 map is that it provides one of the earliest records of the remarkable discoveries made by the Dutch merchant and VOC employee Abel Janszoon Tasman (1603–59). In November 1642, Tasman sighted the island of Tasmania, south of eastern Australia, naming it Van Diemen's Land after the governor of the Dutch East Indies. Tasman went on to land in New Zealand, where he became embroiled in a conflict with local Maori tribes, then returned to Batavia (now Jakarta, the capital of Indonesia), sighting the Fiji islands en route. In a later voyage he mapped much of the north coast of Australia, passing his information to Blaeu, who was responsible for collating new discoveries made by VOC employees.

▶ **Abel Tasman's expedition** was attacked twice by Maoris in December 1642 off the northwest coast of New Zealand's South Island.

6

▲ **THE WEST COAST OF AMERICA AND THE "ISLAND" OF CALIFORNIA** The VOC was interested primarily in trade in the Indian Ocean and the Far East, where its geographical information needed to be particularly good. It was not as involved in the Americas, especially their western coasts, which led to the hesitant and sometimes erroneous geography shown here. Blaeu made the mistake of generations of European cartographers in showing California as a huge island, a belief that remained unchallenged for the next 50 years (see pp.160–61).

Britannia Atlas Road Map

1675 ▪ COPPERPLATE ENGRAVING ▪ 45CM × 51CM (1FT 6IN × 1FT 8IN) ▪ NATIONAL ART LIBRARY,
VICTORIA AND ALBERT MUSEUM, LONDON, UK

SCALE

JOHN OGILBY

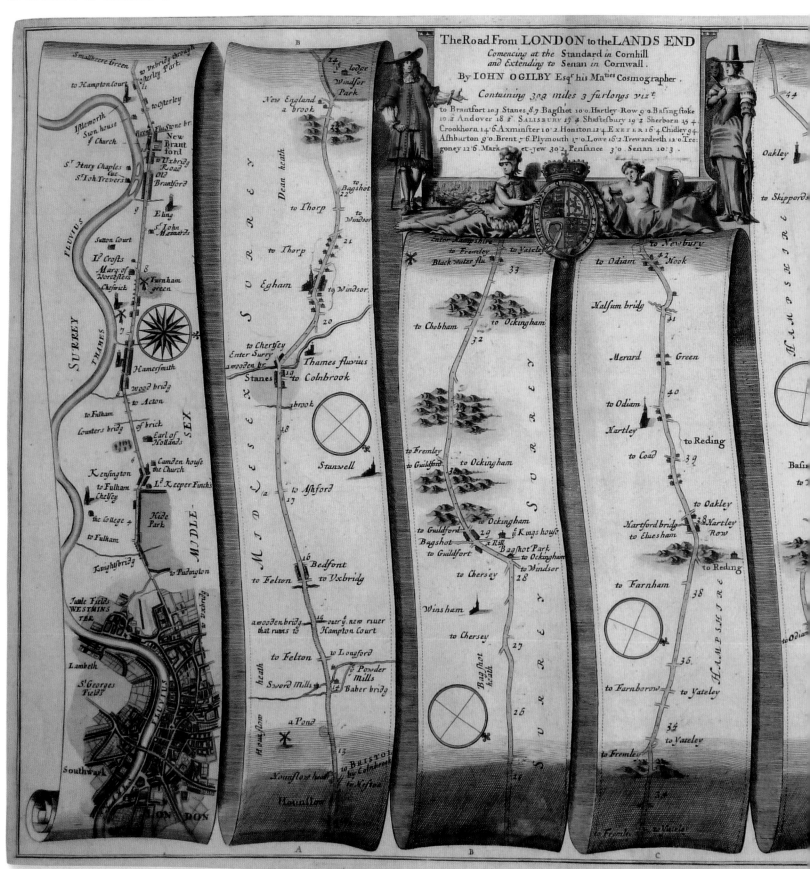

One of the **first national road atlases** to be produced **in western Europe**

JEREMY HARWOOD, BRITISH AUTHOR

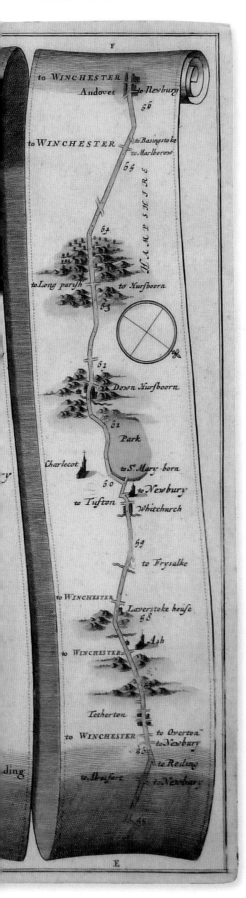

Road maps have a long history, dating all the way back to the Persians and the Romans (see pp.28–31). The first post-medieval road atlas of an entire country to be published in Europe, however, was John Ogilby's *Britannia*, in 1675. It was an enormous, 300-page book, sumptuously engraved and coloured, weighing nearly 7kg (15lb 7oz), and showing the roads of England and Wales. It contained 200 pages of written descriptions and 100 sheets showing 73 major roads, covering more than 12,000km (7,500 miles), although Ogilby claimed his surveyors had mapped nearly four times that figure. Every map used the uniform scale of 2.5cm to 1.6km (1in to 1 mile), which proved so convenient that it was adopted as the Ordnance Survey's standard scale in the 19th century. Each sheet contained a route map covering approximately 110km (70 miles), divided into strips to fit on the page. The sheets mimicked the appearance of handwritten scrolls, creating a *tromp l'oeil* effect, as though the map were a story unfolding as the traveller followed its route.

Despite its decorative flourishes, Ogilby's *Britannia* pursued rigorous standards in surveying and design. The surveyors measured the distances described on each map using hand-pushed wheels, applying basic triangulation methods (see p.165) to ensure relative accuracy. Routes and towns were shown in plan format, with smaller features drawn at an oblique angle. *Britannia* began the trend for standardizing symbols for natural and man-made features such as cultivated land and churches. Higher ground was engraved to show the direction of its incline and steepness. The roads were the map's main focus – although in the 17th century, they were little more than tracks – and changes in direction were indicated by compass roses. *Britannia* was a huge commercial success, although it was more popular among armchair geographers than travellers, due to its size and cost. It remained an indispensable road map until well into the 19th century.

JOHN **OGILBY**

1600-1676

In addition to his cartographic achievements, John Ogilby was also a noted translator and founded the first theatre in Ireland. He began his varied career with an apprenticeship to a dance teacher.

Born in Scotland, Ogilby led a peripatetic life before he became a mapmaker. He was a dance instructor, a teacher, and a theatre impresario in Ireland, where he was caught up in the rebellion of 1641. He moved to England in the late 1640s to work as a translator of the classics. After losing his house in the Great Fire of London (1666), Ogilby was involved in surveying the burnt-out city, leading to a new career in mapmaking. In 1669, he proposed a six-volume geographical description of the world, culminating in *Britannia*. As an enthusiastic royalist, he was appointed Charles II's "Cosmographer and Geographic Printer" in 1674. He died just months after *Britannia*'s publication.

Visual tour

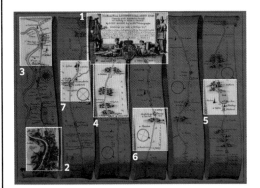

KEY

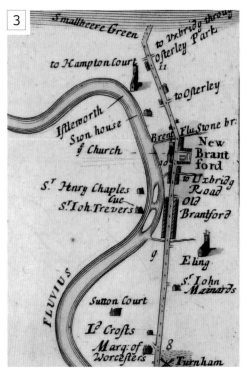

1

▶ **ADVERTISING NEW ROUTES** Restoration figures stand either side of the title, which delineates distances along the route shown below, proudly displaying the English King Charles II's patronage, with Ogilby's official title, and two figures, one reminiscent of Britannia herself, balancing the royal coat of arms over the land.

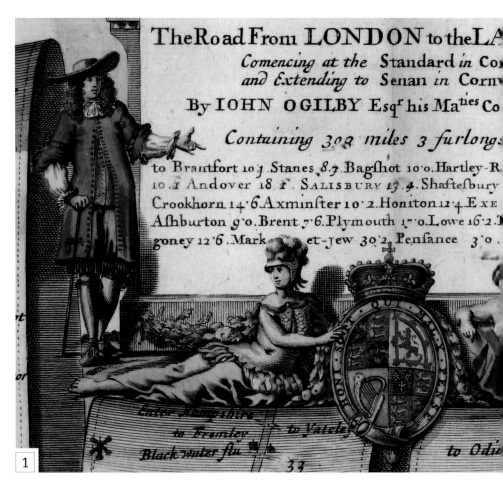

2

▲ **LONDON** The route of this map to Land's End in Cornwall starts in London, with the River Thames, London Bridge, and the southern suburbs of Lambeth and Southwark all drawn in plan. South London is shown on the left. The city is oriented with west at the top to reflect the direction of travel.

▼ **TURNHAM GREEN TO HAMPTON COURT** The route broadly follows road and river through west London's suburbs towards the royal palace of Hampton Court, all shown from above. Road distances are shown in miles, rather like modern guides, and symbols for churches, crossroads, and even barns are rendered as pictorial symbols.

3

4

▲ **OCKINGHAM** Moving into rural areas of Surrey such as Ockingham, Ogilby tries to represent relief through his use of engraved lines. The hills on either side of the road are shown as shallow inclines up and down. All across the map, higher hillsides are drawn to show the direction of their incline.

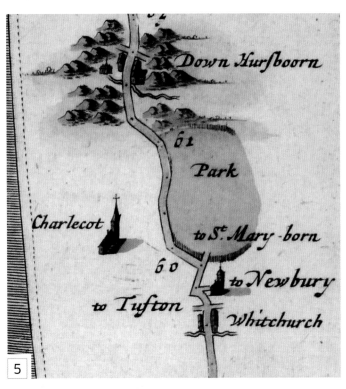

▼ **FARNHAM** The road to Farnham – a notoriously pro-parliamentary stronghold during the English Civil Wars of 1642–51 – has few features apart from the simple red compass rose with a golden *fleur de lys* pointing northwards. Each strip contains a compass rose, although the first one is (for some reason) more decorative than the others. They all indicate the general bearing of that particular stretch of road.

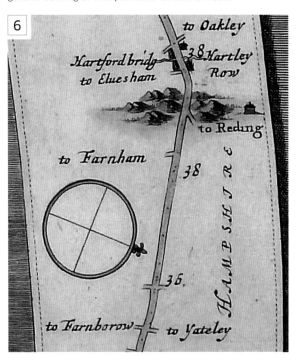

▲ **HURSTBOURNE** Here, Ogilby has drawn the area around Hurstbourne in Hampshire including its nearby enclosed park, shown with a ring of fence palings. There are the usual symbols for streams, buildings, churches, and hilly inclines in the western direction. However, the church to the west labelled "Charlecot" is actually Tufton Church.

▶ **EGHAM TO WINDSOR** Despite his enthusiasm for the recently restored English monarchy, whose arms emblazon the top of the map, Ogilby shows little interest in depicting royal buildings or locations with more prominence than others. Here, the route from humble Egham to the royal castle at Windsor is treated in the same way as any other.

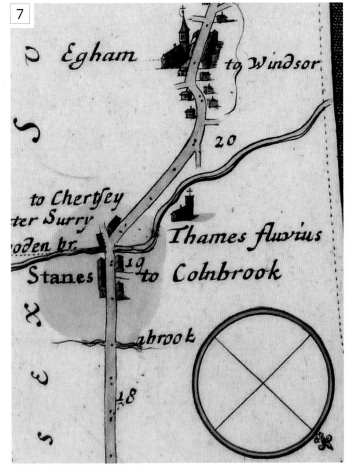

IN **CONTEXT**

Ogilby's *Britannia* was part of a move to map nations and their colonies, with the aim of securing state borders and tightening control over contested territories. Just 10 years before *Britannia*'s publication, the English colonial rulers in Ireland ordered the Down Survey to map the entire country, enabling a massive transfer of land from Irish Catholics to English Protestants. New instruments such as theodolites and wooden poles improved the accuracy of such surveys, and also gave nation states a new means to exercise their power.

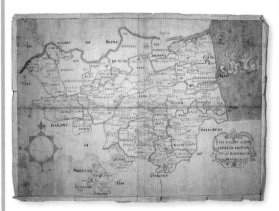

▲ **The Down Survey of Ireland** was carried out between 1656 and 1658 and was the first detailed land survey to take place anywhere in the world.

Map of New England

1677 ▪ WOODCUT ▪ 30CM × 39CM (1FT × 1FT 3¼IN)
▪ MASSACHUSETTS HISTORICAL SOCIETY, BOSTON, USA

JOHN FOSTER

SCALE

This *Map of New England*, attributed to the Boston printer John Foster, has the distinction of being the first map ever to be printed in the Americas. It shows the English colony of New England on the continent's east coast, from Nantucket (bottom left) to Pemaquid Point (bottom right), and from New Haven (top left) inland to the White Hills of New Hampshire (middle right). Today, its orientation seems strange because west, not north, is at the top: this is how America would have been viewed as if seen across the Atlantic from Western Europe. To make sense to a modern eye, the map should be rotated 90 degrees anticlockwise.

The map was made by John Foster to illustrate clergyman William Hubbard's book, *Narrative of the Troubles with the Indians in New-England* (1677). Hubbard described the bloody conflict between the Native American Wampanoag people and the Puritan English settlers of the Massachusetts Bay Colony during King Philip's War (1675–78), named after the Wampanoag chief Metacom (called "King Philip" by the English). The Wampanoag were narrowly defeated; Foster's map quietly records the victory, and draws the colony's northern and southern boundaries, shown by the map's two vertical parallel lines. More than simply a record of geography, on closer inspection this map reveals the story of a colony barely surviving in a hostile environment.

> To anyone with a **taste for American beginnings** it is a fascinating **map of unsurpassed beauty**
>
> **RICHARD B. HOLMAN**, *JOHN FOSTER'S WOODCUT MAP OF NEW ENGLAND*

JOHN **FOSTER**

1648-1681

Known as "The ingenious mathematician and printer", Foster was the earliest American engraver and the first book printer in Boston, where he was born. His pioneering work included some of America's earliest printed literature.

Foster graduated from Harvard University in 1667, then worked as a schoolteacher as well as a doctor, before establishing a printing office in Boston in 1675. He went on to publish various books and images, including the first medical pamphlet on measles, astrological almanacs, Puritan religious tracts, engraved portraits of some of the movement's leading figures (including noted Puritan clergyman the Reverend Richard Mather), topographic views of Boston, and designs for the Massachusetts Bay Company seal. He died at the age of just 32, from tuberculosis.

Visual tour

KEY

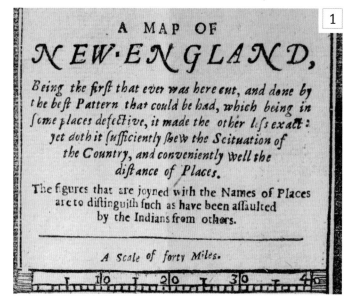

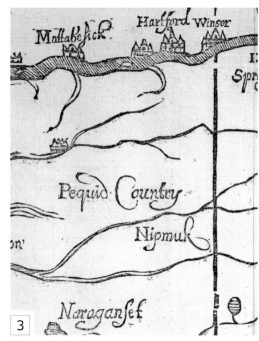

◀ **CARTOUCHE AND SCALE BAR**
Foster deftly mixes science and politics in his title. It contains a scale bar in an attempt to suggest the accurate measurement of the territory. It also advertises itself as the "first" and "best" New England map on the market, showing settlers "the situation of the country", and acknowledging recent conflicts by numbering the places "assaulted" by the Wampanoag.

◀ **THE WHITE HILLS** The map exists in two early states. The first version is shown here; a later, less accurate version was also printed, with this area labelled "The Wine Hills", and various other mistakes. Foster appears to have printed this version in Boston and shipped it back to England, where the second version confused many of the place names.

▲ **PEQUID COUNTRY** On the colony's southern boundary, Foster demarcates indigenous Pequid, Nipnuck, and Narragansett territory (in modern-day Connecticut). Encroaching upon the Native American territories are Puritan settlements with English names such as New London, Windsor, and Northampton, many numbered to indicate Indian attacks. Modern maps show that the English names prevailed, and most local names were expunged within several generations.

▶ **CAPE COD BAY AND COMPASS** The arrival of more settlers from England, blown into the distinctively curled bay on ships and directed by the compass rose pointing north at the bottom of the map, hints at the inevitable eclipse of the indigenous names and ways of life. The distinctive horizontal stippling showing the sea follows the grain of the wood block.

▼ **THE SOUTHERN BOUNDARY** The colony's southern boundary marks a linguistic struggle over territory. Martin's Vineyard (today Martha's Vineyard) was probably named after John Martin, an English sea captain who landed there in 1602. "Nantucket" comes from the Algonquian for "faraway island". Although some local names survived, the natives did not.

▶ **BEYOND THE PALE** North of the dividing line around Casco Bay in Maine, well beyond the colony's territory, are what appear to be armed locals emerging from wooded areas. They are directly above the English settlements numbered as being attacked by Native Americans. This is frontier country, capturing the settler's fear of an alien, inhospitable landscape.

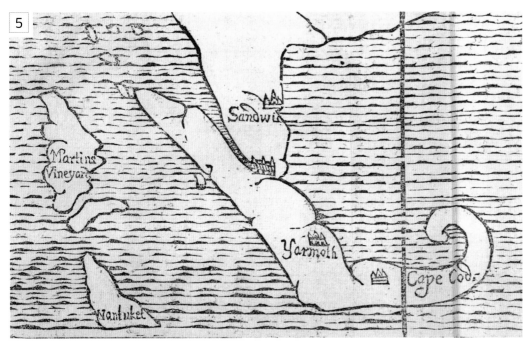

ON **TECHNIQUE**

The *Map of New England* was made using the woodcut printing technique, invented in the mid-15th century. John Foster carved out the non-printed areas (which appear blank on the map) from a plank of wood, using a knife or chisel. This left a stark linear design of the map in relief, which was then inked and printed onto paper.

▲ **A working wood block** printing press today.

Corrected Map of France

1693 ▪ ENGRAVING ▪ 21.6CM × 27.2CM (8½IN × 9½IN) ▪ BIBLIOTHÈQUE NATIONALE, PARIS, FRANCE

SCALE

JEAN PICARD AND PHILIPPE DE LA HIRE

As its title suggests, Jean Picard and Philippe de La Hire's *Corrected Map of France* offered a complete reappraisal of the size and shape of France, based on new astronomical and surveying methods. In 1679, Picard – already known for his innovative use of triangulation and fieldwork techniques for regional surveying – persuaded the French king, Louis XIV, to approve a new map of France that embraced these novel scientific methods. With support from the Observatoire de Paris and the newly formed Académie Royal des Sciences, Picard married his practical fieldwork with the more theoretical and mathematical skills of La Hire. They proved to be an inspired double act that changed the course of European, as well as French, mapmaking.

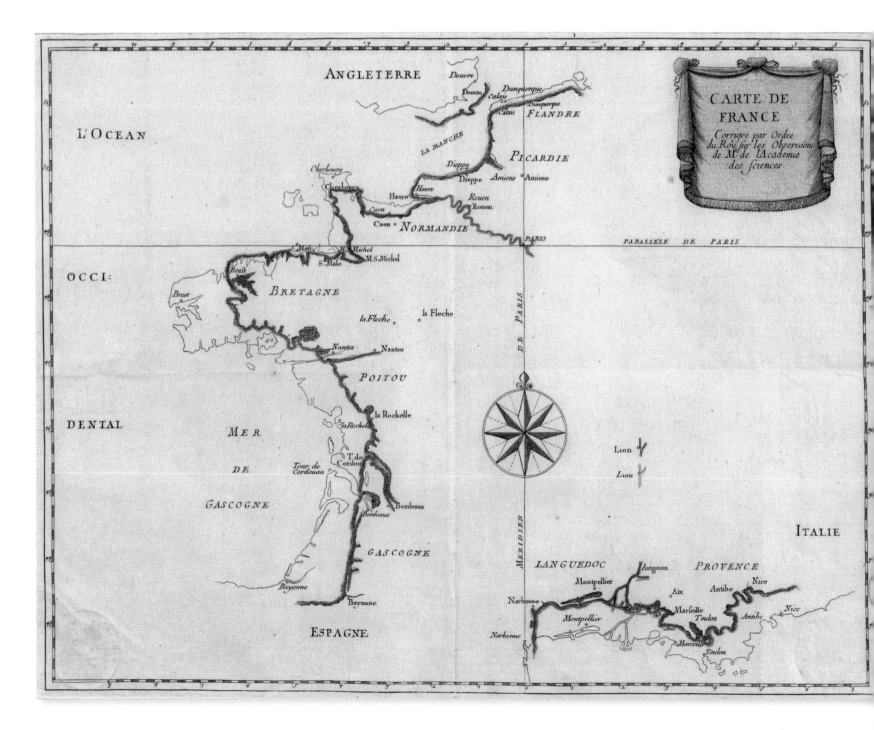

In 1679, Picard and La Hire embarked on a coastal survey that they regarded as a prelude to the first systematic map of France. After mapping the Gascony coast they moved south, then split up to survey the northeast and northwest coasts. Following three years of painstaking work, a manuscript map was drawn just before Picard's death in 1682. Its outline – overlaid with the most accurate extant map, by Nicolas Sanson – revealed that France had an area of 25,386 square leagues: Sanson had put the figure at 31,657. Science had ripped up the old map of France and dramatically reduced its size. King Louis XIV was said to have complained that his geographers had cost him far more territory than any invading army.

JEAN **PICARD** AND PHILIPPE DE **LA HIRE**

1620–1682 AND 1640–1718

Early members of the Académie Royal des Sciences in France, Picard and La Hire were widely regarded as leaders in their respective fields of astronomy and geometry.

A priest, painter, and astronomer, Picard began mapmaking in the 1660s with the support of King Louis XIV. He employed the latest mathematical and fieldwork techniques to recalculate the area of France. La Hire was a theoretical mathematician, writing several astronomical tables on celestial movements and abstract geometry. In 1683, he took the chair in mathematics at the Collège Royale.

Visual tour

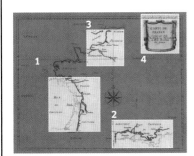

KEY

▶ **THE ATLANTIC COASTLINE** In 1679, Picard and La Hire surveyed the Gascony coast using triangulation and astronomical observations. To their amazement they found that most places along the coast – including the Breton port of Brest, France's major naval base – had previously been plotted 30 leagues too far west, effectively placing them at sea.

1

3

▲ **PARIS MERIDIAN** The meridian running through Paris was accurately mapped for the very first time. The line had been recently recalculated by Picard, and the Observatoire de Paris was designed to bisect it.

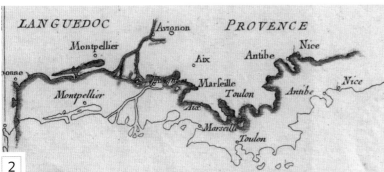

2

▲ **THE SOUTHERN COAST – LANGUEDOC AND PROVENCE** Some of the final measurements were taken in 1682 along the strategically important Mediterranean coastline, from Montpellier in the west to Nice in the east. Picard confirmed many of his earlier findings: the country's size had been accidentally overestimated. As a result, the coast was completely redrawn, retreating northwards and shrinking the country even further.

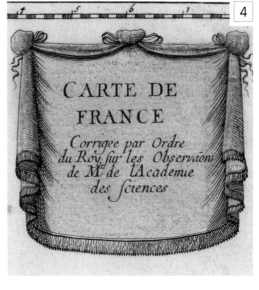

4

◀ **CARTOUCHE** The map's *trompe l'oeil* cartouche containing its title explains that the "corrections" were by order of the king, and undertaken by the respected Académie Royal des Sciences – probably to mitigate any potential political damage caused by presenting such a shocking reduction in France's size.

Map of the Holy Land

1695 ■ ENGRAVING ■ 26CM × 48CM (10¼IN × 1FT 6IN) ■ LIBRARY OF CONGRESS, WASHINGTON, DC, USA

ABRAHAM BAR-JACOB

SCALE

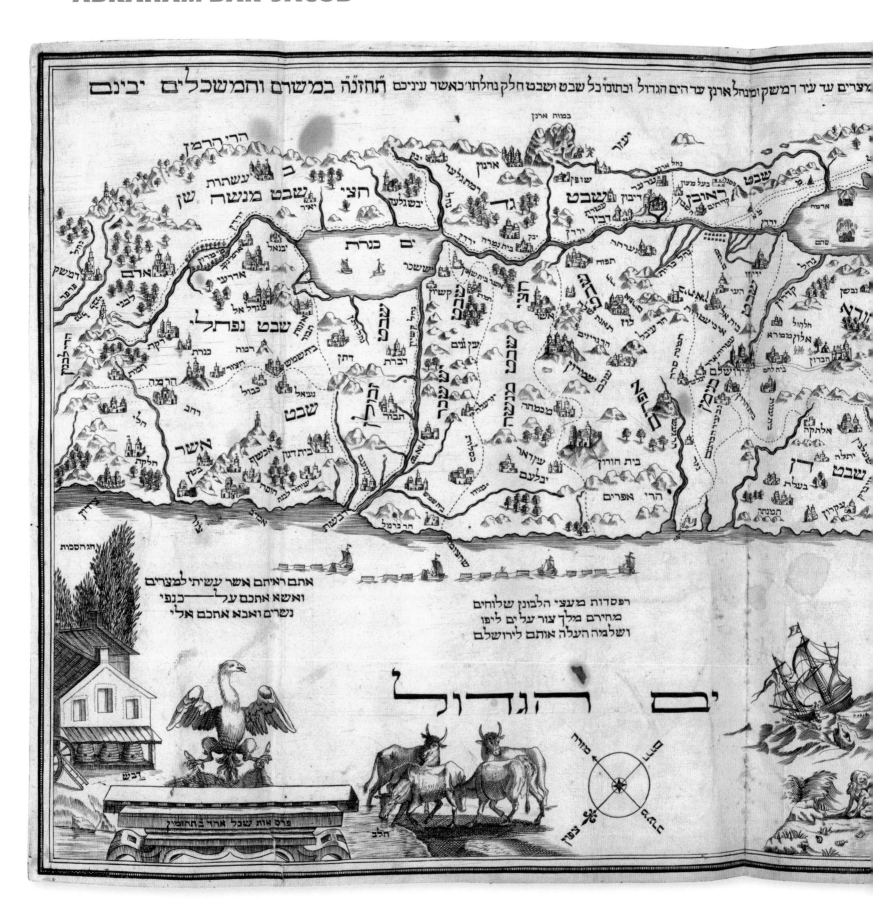

The cartographer, presumably a proselyte, not only **translated the map into Hebrew** but turned it into a Jewish map by **removing the references to Christianity and the New Testament**

"

ARIEL TISHBY, *HOLY LAND IN MAPS*

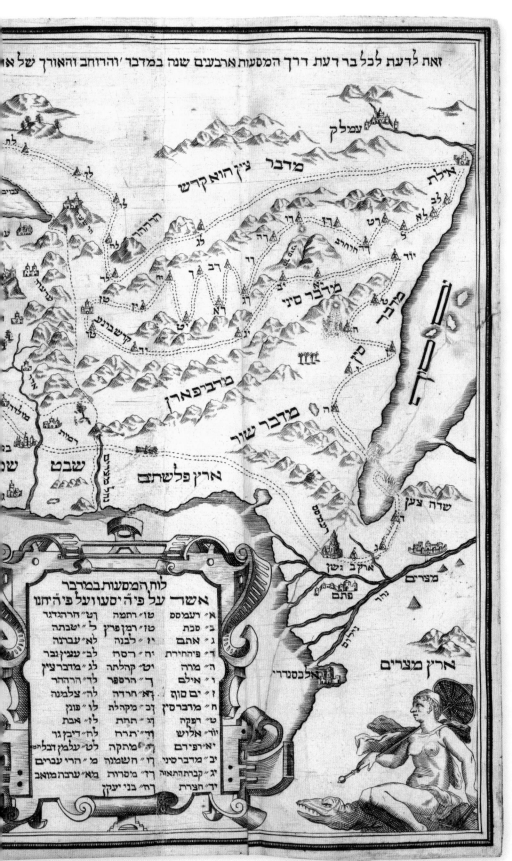

This is one of the earliest maps of the Holy Land created in Hebrew, and was engraved in Amsterdam by Abraham Bar-Jacob, a convert to Judaism from Christianity. It shows in exquisite detail the story of the Israelites' Exodus from Egypt, the arrival in the Promised Land of Canaan, and the designation of the Twelve Tribes of Israel's territories. It has a very specific function as an illustration for a *Haggadah*, the prayer book used in the ritual celebration of Passover, which commemorates the Exodus from Egypt. The map appeared in the famous *Amsterdam Haggadah*, published by Moses Wesel. In contrast to many Christian maps of the region, this is oriented with southeast at the top, with the eastern Mediterranean running along the bottom. Stretching from the Nile on the right to Damascus, it is an enduring celebration of the foundation of Israel.

ABRAHAM **BAR-JACOB**

C.LATE 17TH–EARLY 18TH CENTURY

Very little is known about the map's maker, as Abraham Bar-Jacob took this name after his conversion to Judaism from Christianity.

Bar-Jacob grew up in the German Rhineland where he trained as a pastor before converting to Judaism. He then moved to Amsterdam in the Netherlands and joined its thriving Jewish community, where he worked within its tolerant, multilingual publishing industry, engraving various mystical religious images and producing writings on Judaism.

Visual tour

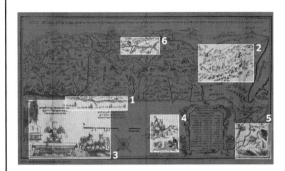

KEY

▼ **BUILDING A NATION** Just off the southern coast, ships tow the cedars of Lebanon, sent by King Hiram of Tyre to King Solomon (C.970–931 BCE). These enabled him to build the Holy Temple on Mount Zion – a symbol of the founding of Israel.

▼ **EXODUS** Across the Sinai desert, the map traces the Children of Israel's Exodus from Egypt and journey to the Promised Land. Their wandering is shown in dotted parallel lines, with each encampment marked by a number keyed to the cartouche below.

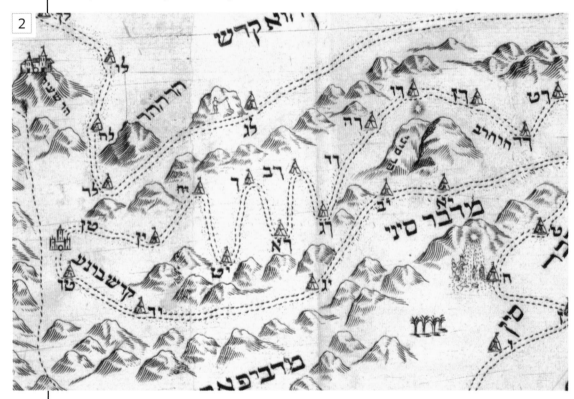

▶ **SACRED SYMBOLISM** This scene represents the Promised Land's virtues. The house with beehives and the cattle to the right symbolize the fertile future of Israel as a land of milk and honey. The eagle refers to God's Covenant with Moses, and how God says in the Bible, "I bore you on eagle's wings" through the Egyptian wilderness (Exodus, 19:4).

IN CONTEXT

The Holy Land had been shown on Christian maps since medieval times. However, these maps' abiding interest was in not just the Old Testament stories, but also those of the New Testament and of Christ's life, death, and resurrection. The 16th century witnessed a marked increase in these maps as Catholics, Protestants, and Jews asserted their theological and visual interpretations of the Holy Land. Catholic maps traditionally tended to focus on Christ's life; Protestants emphasized the Apostles; the later Hebraic tradition focused on the Old Testament foundation of Israel, in particular the story of the Exodus.

▲ **This 16th-century map of the Holy Land** featured in the world's first atlas, *Theatrum Orbis Terrarum* by Abraham Ortelius, a Catholic.

◄ **JONAH AND THE FISH**
Jonah was a Jewish prophet who tried to escape God's commandment to preach against the city of Nineveh. He is shown here being swallowed by a great fish, but saved through divine intervention. The story emphasizes that God's plan is irrevocable, and that he will rescue the Israelites whenever they face mortal adversity.

◄ **NILE CROCODILE**
The sinful temptations of Egypt are represented on the map by a naked woman riding a Nile crocodile, holding a parasol. It may even be an oblique reference to the discovery of the baby Moses by the Pharaoh's daughter while bathing in the Nile.

▼ **REACHING THE PROMISED LAND**
The crossing of the River Jordan – the culmination of 40 years of wandering – and the arrival into the Promised Land, is marked by the 12 stones laid on the river's west bank, representing the Twelve Tribes of Israel.

Land Passage to California

1710 ■ ENGRAVING ■ 36.8CM × 24.1CM (1FT 2½IN × 9½IN) ■ LIBRARY OF CONGRESS, WASHINGTON, DC, USA

SCALE

EUSEBIO FRANCISCO KINO

Mapmaking can often reproduce errors as truth until some intrepid figure challenges them, as in the case of the Jesuit missionary and mapmaker, Eusebio Francisco Kino. Since the early Spanish discoveries made across the Americas throughout the 16th century, California had been projected as an enormous island. An imaginary stretch of water, the "Strait of Anian", was often shown separating modern-day Arizona from California, even by respected mapmakers such as Ortelius and Blaeu (see pp.142–45). This erroneous view was perpetuated by the difficulty of exploring the Gulf of California by sea, and the inhospitable, arid desert by land.

A bold missionary

The only person with sufficient determination to explore the region was a missionary, Kino, who made a series of arduous journeys across northern Mexico in the 1680s to establish Jesuit missions. He began to map the area, becoming convinced that, as he wrote in one of his reports, "*California no es isla*" ("California is not an island"). In a series of maps drawn in the mid-1680s, Kino showed that Baja California was connected to the American mainland at the mouth of the Colorado River. However, because he was not a trained mapmaker, his claims were not accepted by the Spanish authorities until the 1740s – three decades after Kino's death.

Visual tour

KEY

1

▲ **THE COLORADO RIVER** Kino traversed the mouth of the Colorado River, proving that the Colorado River ("Coloratus") connected California to the mainland.

2

◀ **"MARE DE LA CALIFORNIA"** For the first time ever Kino shows Baja California as a peninsula with a gulf around 240km (150 miles) wide connecting it to the Mexican mainland.

3

▲ **PIMERIA AND SONORA** The map also provides a detailed record of Kino's travels and missionary work across the arid, rugged terrain of Pimeria Alta, one of North America's hottest deserts. It includes the mission established at Populo.

EUSEBIO FRANCISCO **KINO**

1645–1711

Born in Italy and trained as a Jesuit in Bavaria, Kino received his holy orders in 1677 and was sent to the Americas in 1683, by which time he was already developing an interest in astronomy and mapmaking.

Sometimes known as Eusebio "Chino" or "Chini", Kino conducted his missionary work in what is now California, Arizona, and Mexico, particularly the area known as Pimeria Alta. Working closely with the indigenous population, he established 24 religious missions in the region, converting locals, mapping the land, and encouraging the use of new agricultural methods. Kino was a vociferous critic of Spanish colonial rule, attacking the use of slaves in local silver mines. He died of a fever while undertaking missionary work in northwest Mexico.

TABULA CALIFORNIÆ Anno 1702.

Ex autoptica observatione delineata á R.P.Chino e S.I.

Via terrestris in Californiam comperta et detecta Per R.Patrem Eusebium Fran.Chino e S.I. Germanum. Adnotatis novis Missionibus ejusdem Soctis ab Anno 1698. ad annum 1701.

Miliaria Gallica.

Tabula Geographica R.P.Eusebij Franc: Kino Tridentini e Soc. IESU.

Tabula Chartæ Patris Chino addita.

Annotatio.

Pars hujus Tabulæ A.B.C.D. e Chartâ Topographicâ R.P.Eusebij Francisci Chino fuit transumpta, appendix autem C.D.E.F. e tabulis antiquioribus est adjecta. Gradus latitudinis cum eodem autore designavimus, longitudinis vero illius exemplo omisimus.

Zu Num. 53. in dem II. Theil und zu Num. 13. in dem III. Theil.

New Map of France

1744 ■ ENGRAVING ■ 58CM × 91CM (1FT 11IN × 3FT) ■ BIBLIOTHÈQUE NATIONALE, PARIS, FRANCE

SCALE

CÉSAR-FRANÇOIS CASSINI DE THURY

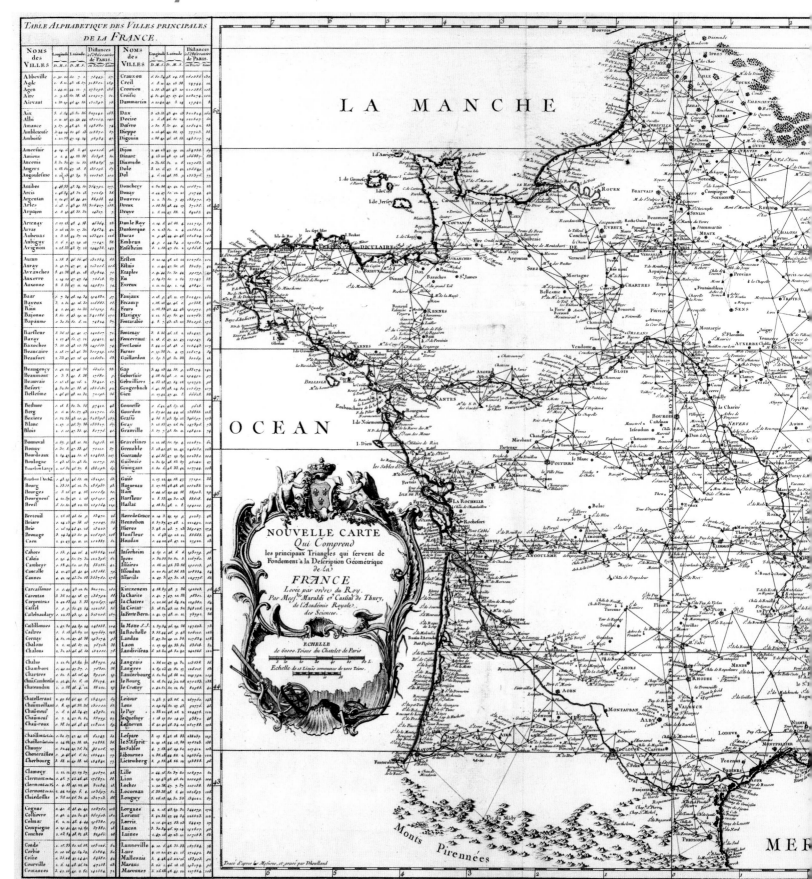

The French launched the first organized attempts to map their nation using new scientific techniques – surveys that used triangulation and various instruments to produce maps of unparalleled detail and accuracy. From the late 17th century, a noted scientific family, the Cassinis, was given the task of surveying the entire kingdom by King Louis XIV, for the purposes of military defence and more efficient taxation. The first survey began in 1733, led by teams of surveyors who fanned out across the country with cumbersome measuring equipment, sending back their findings to the Cassini family members working at the Observatoire de Paris. Using the latest methods for accurately triangulating distances, first Jacques Cassini and then his son César-François Cassini de Thury oversaw a vast publishing project to map the nation in all its detail.

In 1744, the survey was completed and ready for publication. The *New Map of France* (also known simply as *Carte de Cassini*) was sumptuously engraved on 18 sheets on a scale of 1:1,800,000. It shows the country as a network of tightly connected triangles based on the surveyor's new methods, each one carefully measured and calculated over years of fieldwork. It is composed of an amazing 800 triangles and 19 base lines.

However, despite the map's unprecedented level of accuracy and detail, there are also noticeable shortfalls. The physical contours of the land are almost completely absent, as Cassini de Thury had no way of measuring and showing altitude and elevation. Upon publication he admitted, "We haven't visited each farm or followed and measured the course of every river". Large mountainous areas such as the Pyrenees, the Alps, and the Massif Central are simply left blank.

This was, however, a first geometrical step towards the completion of a standardized map of a nation. Cassini de Thury's method and style of regimented lettering, symbols and physical features laid the foundation for all subsequent surveys.

CÉSAR-FRANÇOIS **CASSINI DE THURY**

1714–1784

Originally from Italy, the Cassini family relocated to France in the mid-17th century. There, they dominated the fields of astrology and cartography for well over a century.

César-François Cassini de Thury took over the national surveying projects started by his father Jacques Cassini. César-François also inherited many of his father's titles as well as his projects, including membership of the Académie Royal des Sciences, directorship of the Observatoire de Paris, and the publication of the first stage of a monumental national survey in 1744.

Visual tour

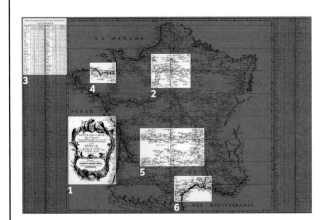

KEY

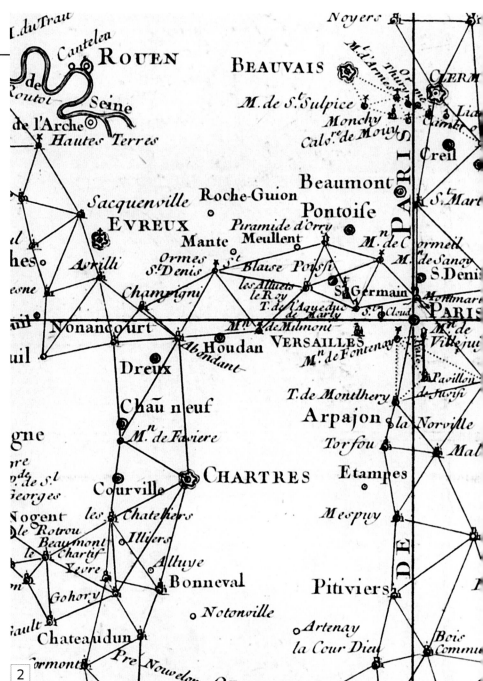

2

▲ **PARIS** At the centre of a series of converging triangles stands Paris. The Paris meridian runs directly through both the city and the famous observatory. It is bisected by a perpendicular line from which all the triangles measured by trigonometry originated. On Cassini de Thury's maps, all lines, rather than roads, lead to Paris.

▲ **ART, SCIENCE, AND METHOD** The beautifully engraved title announces the map's striking modernity and its new methods – triangulation and geometry – which are endorsed by both the Académie Royal des Sciences and King Louis XIV. The standard scale is prominently displayed, as are the tools of the cartographic trade. The word "France" is emphasized, leaving no doubt as to the map's patriotic intent.

▶ **ALPHABETICAL TABLE** Cassini de Thury's maps were full of innovations, including alphabetical tables listing key towns and cities. Their latitude and longitude are given – based on surveying methods and astronomical observations – as well as the distances to the map's centre, Paris.

TABLE ALPHABETIQUE DES VILLES PRINCIPALES DE LA FRANCE.

NOMS des VILLES	Longitude D.M.S.	Latitude D.M.S.	Distances a l'Observatoire de PARIS. en Toises Lineai	NOMS des VILLES	Longitude D.M.S.	Latitude D.M.S.	Distances a l'Observatoire de PARIS. en Toises Lineai
Abbeville	0. 30. 22	50. 7. 1	72443.	47.	Crauzon	6. 50.34	48. 14.58	160816.	130.
Agde	1. 8.11	43. 18.59	318801.	189	Creil	0. 8.16	49. 16.38.	24744	10.
Agen	1. 44.41	44. 12. 7	258050.	180.	Cremieu	2. 55.18	45. 43. 20	202382.	105
Aire	0. 3.18	50. 38. 18.	104917.	94.	Croisic	4. 50.40	47. 17. 40	202474.	102.
Airvaut	1. 38.49	46. 49. 28.	161398.	78.	Dammartin	0. 20.43	49. 3. 14	17344.	8.
Aix	3. 6.34	43. 31. 36	347142.	168	Dax	3. 13.08	43. 42. 22	321824.	160.
Albi	0. 11. 20	43. 55. 44	280102.	140.	Decise	1. 8.18	46. 50. 24	121807.	60.
Amance	3. 87. 30	48. 45. 8	143080.	74.	Daivre	0. 30. 43	50. 40	108142.	83.
Ambleteuse	8. 44.12	50. 45. 12.	118570.	87.	Dieppe	1. 16.48	49. 55. 88.	77713.	38.
Amboise	1. 00.53	47. 24. 24	90384.	47.	Digouin	1. 38.40	46. 28. 82.	148307.	74.
Amerfuir	4. 24. 6	48. 8. 41	190208.	92.	Dijon	2. 43.08	47. 19. 21	134038.	67.
Amiens	0. 2. 4	49. 53. 33.	60348.	30.	Dinant	4. 23.40	48. 27. 16	166387.	82.
Aucenis	3. 31.34	47. 21. 50	188483.	70.	Dixmude	0. 31.36	51. 2. 6	127082.	63.
Angers	2. 03.24	47. 28. 8	188008.	82	Dole	3. 10.18	47. 5. 44	158840.	79.
Angouleine	1. 08.40	45. 38. 3.	220843.	109	Dol	4. 6.18	48. 32. 52	181348.	86.
Antibes	4. 48.53	43. 34. 50	354790.	177.	Donchery	2. 34.58	49. 41. 52	108871.	52.
Arcis	1. 48.34	48. 31. 12	70268.	36.	Douay	0. 44.47	50. 21. 52	94744.	44.
Argentan	2. 24.28	48. 44. 49	88068.	44.	Douvres	1. 2. 34	50. 8. 13	188270.	93.
Arles	2. 43. 0	43. 40. 33	308402.	103.	Dreux	0. 58.36	48. 44. 13	37188.	17.
Arpajon	0. 8.40	48. 35. 32	14847.	7.	Druye	1. 8.44. 47.	53. 12	84083.	41.
Artenay	0. 47.48	48. 4. 68	46364.	23	Dun le Roy	0. 14. 6	46. 58. 6.	111748.	68.
Arras	0. 05.10	50. 17. 50	84624.	41.	Dunkerque	0. 22.51	51. 2. 4	125810.	62.
Aubenas	1. 3.42	44. 37. 22	283491.	106.	Duras	1. 40	44. 48. 4	188270.	93.
Anbiony	1. 8. 7	47. 49. 12	184280.	92.	Embrun	5. 13. 44. 34	4	138668.	68.
…on	2. 28.53	43. 37. 19	294933.	147.	Eaulsheim	5. 12.32	47. 47. 25	198608.	98.
	1. 33. 8	46. 36. 88.	131884.	66.	Eriten	8. 20.14	48. 28.	202776.	101.

3

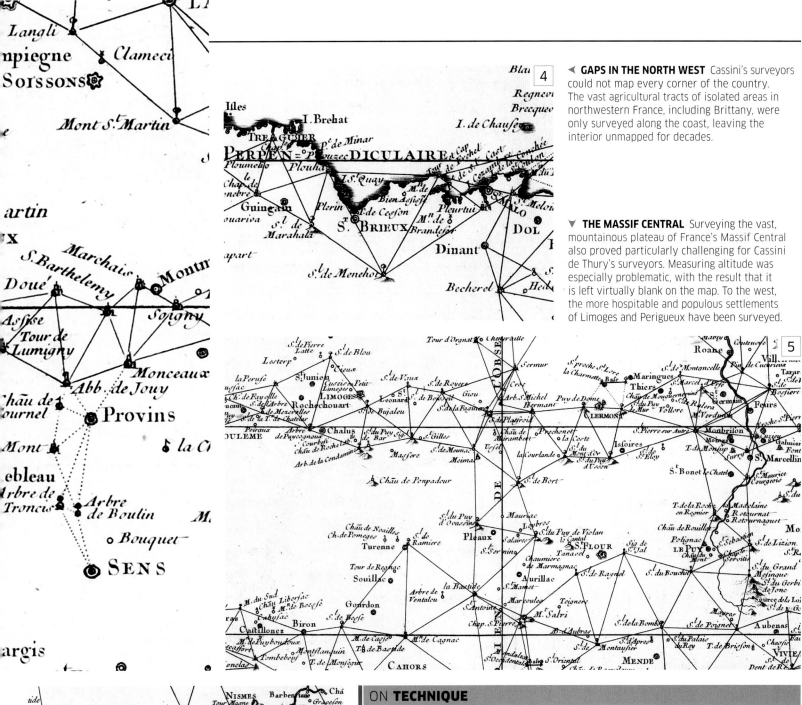

◀ **GAPS IN THE NORTH WEST** Cassini's surveyors could not map every corner of the country. The vast agricultural tracts of isolated areas in northwestern France, including Brittany, were only surveyed along the coast, leaving the interior unmapped for decades.

▼ **THE MASSIF CENTRAL** Surveying the vast, mountainous plateau of France's Massif Central also proved particularly challenging for Cassini de Thury's surveyors. Measuring altitude was especially problematic, with the result that it is left virtually blank on the map. To the west, the more hospitable and populous settlements of Limoges and Perigueux have been surveyed.

▲ **MONTPELLIER** Interlocking triangles march across the country. One starts near Montpellier, on France's southern coast, making its way up towards Paris. Again, much of the surrounding area remains free of such surveying methods, blank space which seems to have fallen off the national consciousness. Not everyone benefitted from Cassini's maps.

ON **TECHNIQUE**

The Cassini surveys were the first systematic attempts to map a nation using the new technique of triangulation. This involved the exact measurement of a base line, using wooden rods several metres in length. Knowing the precise length between two points, a third could then be identified, often from an elevated position such as a church tower, from which a triangle could then be drawn. Using trigonometrical tables, the distance of the third point from the others could be measured with precision. This process was then repeated across a city, region, and finally the entire country, a method perfected by Cassini's army of surveyors over many years.

▲ **In this woodcut**, a surveyor holding a quadrant measures the height of a church tower using triangulation.

THEMATIC MAPS

- Jain Cosmological Map

- A Map of the British Colonies in North America

- Indian World Map

- Map of All Under Heaven

- A Delineation of the Strata of England and Wales with Part of Scotland

- Japan, Hokkaido to Kyushu

- "Indian Territory" Map

- John Snow's Cholera Map

- Slave Population of the Southern States of the US

- Dr Livingstone's Map of Africa

- Missionary Map

- Descriptive Map of London Poverty, 1898–9

- Marshall Islands Stick Chart

1750–1900

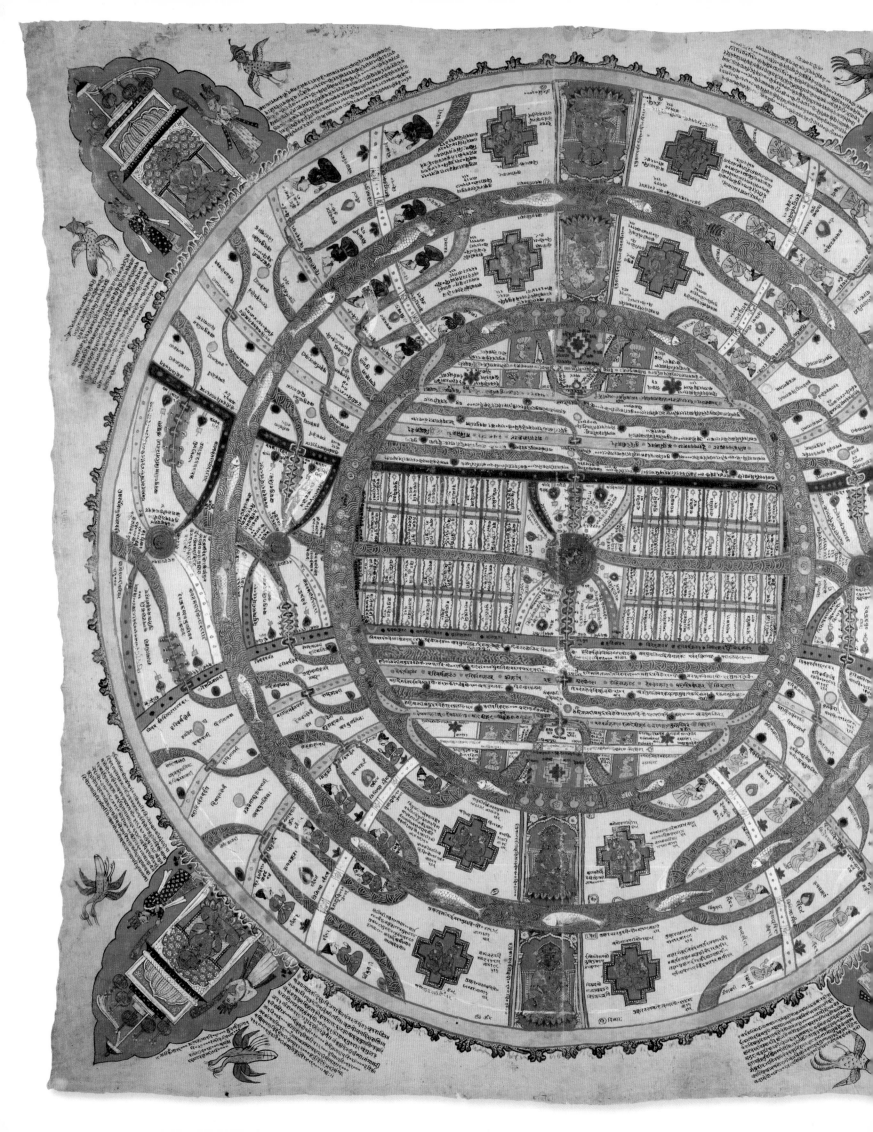

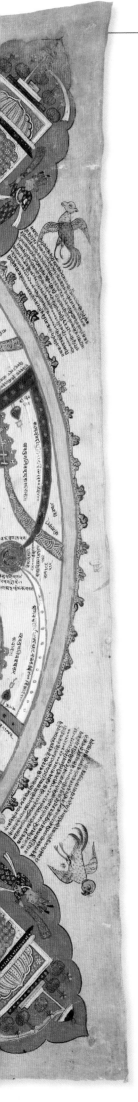

Jain Cosmological Map

1750 ■ PARCHMENT ■ 90CM × 87CM (2FT 11½IN × 2FT 10¼IN) ■ BRITISH LIBRARY, LONDON, UK

SCALE

UNKNOWN

Throughout history, maps have offered powerful and convincing graphic visualizations of various theological world views. None of these are more beautiful than this map illustrating the complex cosmology of Jainism – one of the oldest religions in the world, originating in ancient India. Jains do not believe in creation by a god, claiming instead that *Loka* (the universe) is eternal and infinite; nothing within it is either created or destroyed. In Jain belief *Lokakasa* (cosmic space) is divided vertically into three parts: the *Urdvha Loka* (the Upper World) of divine creatures, split into different "abodes"; the *Madhya Loka* (the Middle World), inhabited by humans and animals; and *Adho Loka* (the Lower World), consisting of seven subterranean hells. Beyond these three regions is *Alokakasa*, a void outside of space and time.

A human world

Most surviving Jain maps, as in the 18th-century example shown here, only represent *Madhya Loka*, the Middle World of humanity, minus the Upper and Lower worlds, effectively providing a cross-section of the terrestrial universe. The mythical, holy mountain of Meru lies at the centre of terrestrial space, in the middle of Jambudvipa – literally the "land of the Jambu (blackberry) trees" – a circular continent inhabited by mankind. This measures 100,000 *yojans* (a Jain unit of measurement equal to around 8–10km or 5–6 miles), or around 725,000km (450,000 miles). The continent is divided into *kshetras* (regions), including Bharata Kshetra (India). Beyond Jambudvipa are seven concentric oceans and continents (of which only two are shown here). One final ocean known as Swayambhu Raman encloses the entire circular middle world.

The sacred Jain cosmographic texts, called *samgrahanis* (compilations), are often illustrated on cloth or paper and displayed in temples. They are used for didactic purposes to show believers the complex nature of their world view. As Jains also believe in reincarnation, such graphic maps allow the faithful to anticipate how they might transmigrate (be reborn) across time.

> Jains envisage our own universe as consisting of a series of netherworlds increasing regularly in size with distance below the world of man and a series of heavenly realms above it

JOSEPH E. SCHWARTZBERG, AMERICAN GEOGRAPHER AND HISTORIAN

Visual tour

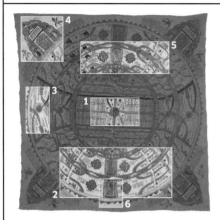

KEY

▶ **MOUNT MERU** At the map's heart lies Mount Meru, the centre of the universe, the holiest mountain in Jainism. It probably draws its inspiration from the Pamir Mountains in central Asia. The four tusk-like protrusions spiralling outwards are mountain ranges called Vidutprabha, Gandhamandana, Malyavata, and Saumanasa. The surrounding area is composed of 16 *videhas* (rectangular provinces), in which mankind dwells.

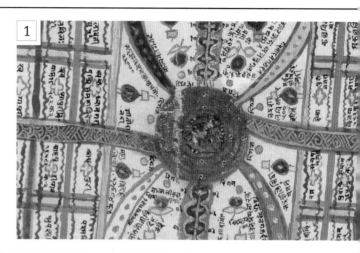

1

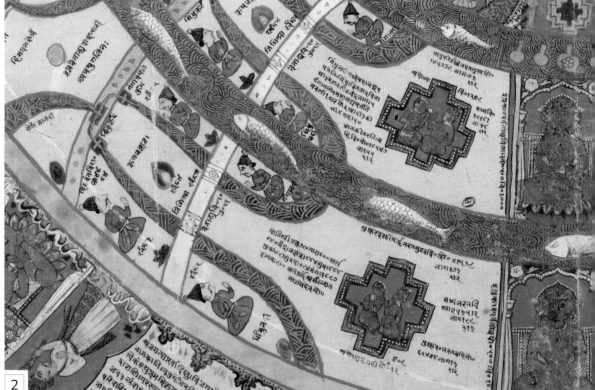

2

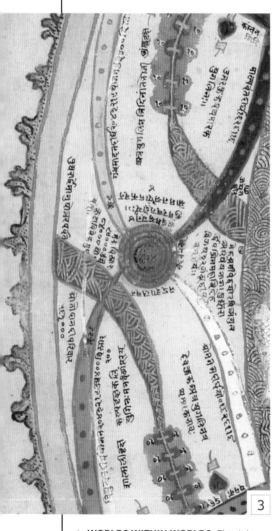

▲ **WORLDS WITHIN WORLDS** The Jain universe is a space of infinite plurality and aesthetic symmetry. Here on the continent of Puskaradvipa (Lotus Islands), named after the flower of purity and renewal that fills the region, another Mount Meru appears. Barely half of this region is populated by mankind, in contrast to the central continents of Jambudvipa and Dhatakikhanda. The outer edge's blue wavy band represents mankind's limits – the mountain range of Manusottara.

3

▲ **BHARATA, OR INDIA** South of Mount Meru is Bharata (India). Its central bell-shaped section is split by two rivers: the Indus to the left, running into the ocean; and the Ganges to the right. The northern areas are the lands of "impure" castes inhabited by *mlecca* (barbarians). The central southern areas are the *aryakhanda* (noble lands), the pure realm of the Aryan castes. This is an area in which humans have to work, and where change and transformation is possible.

▶ **ESOTERIC SYMBOLS** The map is full of religious symbols, some difficult even for Jains to decipher. Beyond the Manusottara mountain range, a *Jina* (see opposite) is surrounded by objects of devotion. He sits beneath a canopy, a symbol of spiritual sovereignty, while heavenly attendants carry fly-whisks, representing divinity. Above, *kalashas* (metal pots) can be seen, symbols of abundance, while the birds with human faces represent the transformation between human and animal.

4

▶ **HUMANITY'S LIMITS** In Jain cosmology, mankind exists within two *kshetras* (regions), represented as concentric islands around Mount Meru. Known as *Manusya Loka* (Human Universe), this world's borders are full of the hope of spiritual liberation, with figures transmigrating from one region to another. In this space, individuals and couples seek harmony, overseen by *Jinas* (see below). The fish demarcate the oceans, but for Jains they also represent boundlessness and the ability to migrate from one place to another.

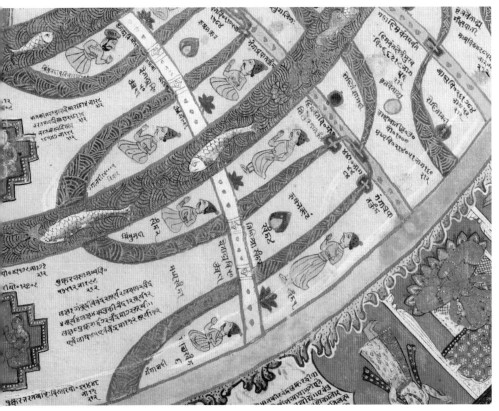

IN CONTEXT

Traditionally, Jain scholars believed that the universe was broad at the top and bottom and narrow in the middle. In fact, Jain cosmology was often depicted as a human body, shown standing upright with arms akimbo. The figure – which could either be a man or a woman – embodied the Jain's three separate *loka* (worlds): the head and torso indicated the heavenly *Urdvha Loka* (Upper World), the hands position on the hips related to the human *Madhya Loka* (Middle World) of Jambudvipa, and the legs symbolized the hellish *Adho Loka* (Lower World).

◀ **JINAS** On the borders of the human world sit the *Jinas* (conquerors or liberators). They are also known as *Tirthankaras* (ford-makers) as they are considered divine guides and teachers showing the faithful the path to liberation. They are immediately recognizable here with their elongated earlobes, meditating under a canopy in the *padmasana* (lotus position) of absolute passivity, displaying the *srivatsa* (chest marking) that distinguishes them from the Buddha.

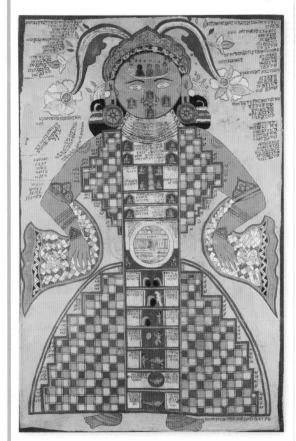

▲ **The Jain tradition of representing** cosmology as a human figure existed alongside more recognizable, circular cosmological maps of the universe.

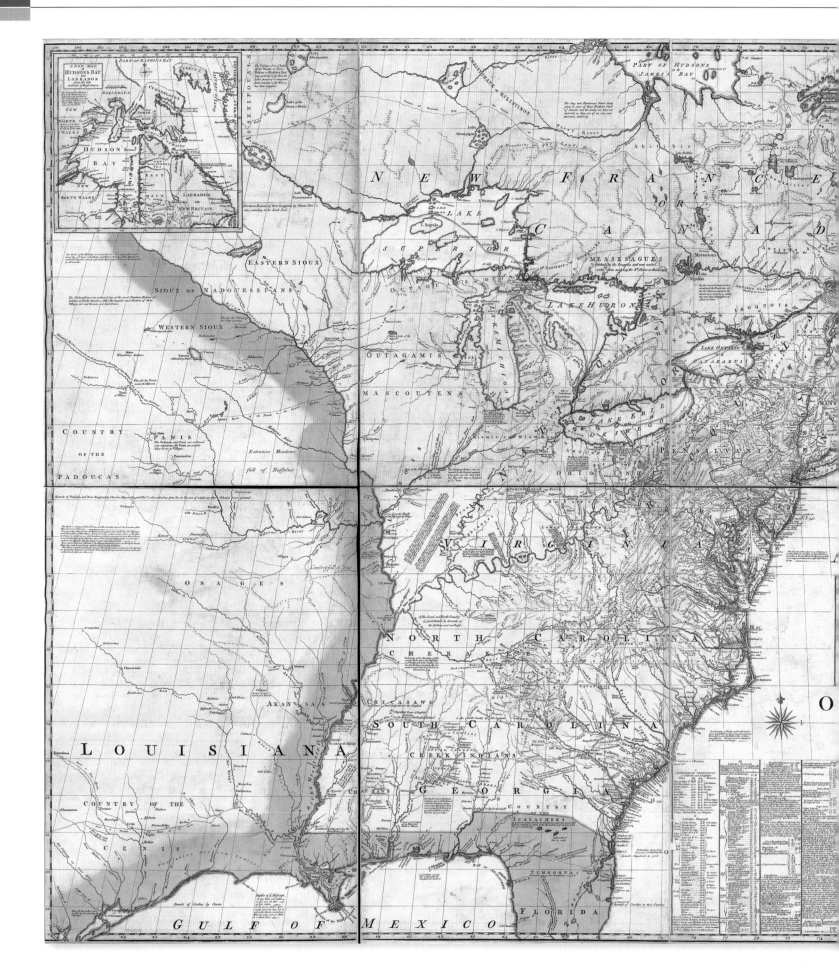

The most important map in North American history

MATTHEW EDNEY, BRITISH MAP HISTORIAN

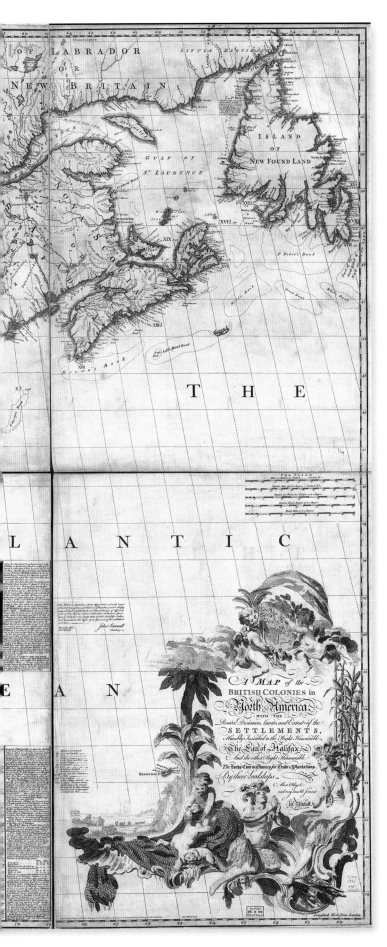

A Map of the British Colonies in North America

1755 ▪ ENGRAVING ▪ 1.36M × 1.95M (4FT 5½IN × 6FT 4¾IN) ▪ LIBRARY OF CONGRESS, WASHINGTON, DC, USA

SCALE

JOHN MITCHELL

The most significant map in the history of the United States was designed by a doctor and botanist, John Mitchell, to chart Britain's North American colonial territories, and to help defend them from French encroachment. The map was supported by George Dunk, 2nd Earl of Halifax and President of the British Board of Trade and Plantations with responsibility for Britain's overseas colonies, to whom the map is dedicated. Both Mitchell and Dunk were concerned that the boundaries agreed between Anglo-French colonial territories under the Treaty of Utrecht (1713–14) were being eroded by French settlers. Halifax's data enabled Mitchell to produce a map of unparalleled detail, showing British dominions in blue-red (with the French in green-yellow) stretching from the Atlantic to beyond the Mississippi, and from the Great Lakes to the Mexican Gulf.

Ironically, although the map was created to demarcate the regions of America under British control, it was subsequently used to agree the boundaries of the newly independent United States at the Treaty of Paris in 1783, at the end of the American Revolutionary War. It was even used to help settle Maine fishery disputes as recently as the 1980s.

JOHN MITCHELL

1711–1768

Born in Virginia, America, and educated in medicine at Edinburgh University in Scotland, John Mitchell was an enthusiastic and respected botanist and student of natural history.

Mitchell returned to his homeland and set up a medical practice at Urbanna, Virginia, in 1735, but moved back across the Atlantic from America to London in the 1740s to seek treatment for both his own and his wife's ill health. After settling in London, he became a member of the Royal Society, met the Earl of Halifax, and began work on his famous map.

Visual tour

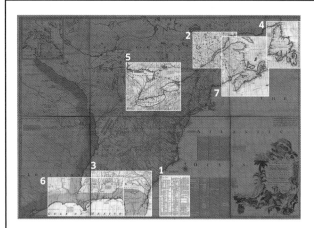

KEY

▼ **LEGEND** Mitchell's legend describes how he made the map. He gives details of the astronomical observations he used to calculate latitude and longitude, and so to produce some of the most accurate figures ever recorded. As he was not a trained cartographer, Mitchell explains that he had synthesized various sources of astronomical data, coastal manuscripts, maps, and local surveys that were available at the time, but which had never before been combined into one map.

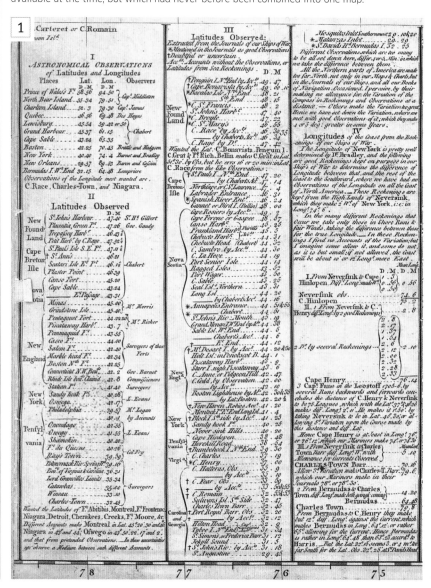

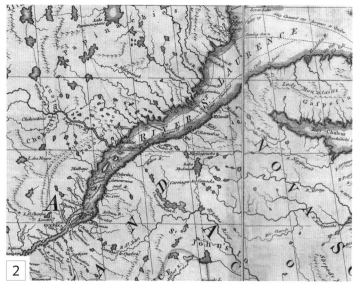

▲ **ST LAWRENCE RIVER** One of the map's most contentious regions was the boundary between New England and the French colony of Québec (bottom left). The British claimed all land to the north, up to and including the strategically important St Lawrence River, an area that contained a large concentration of French settlers. The Treaty of Utrecht left the boundary unclear, leading to confusion and conflict between the two sides. Mitchell's boundary asserts quite clearly that this is British territory.

6 GULF OF MEXICO

▲ **GULF OF MEXICO** The mouth of the Mississippi and the Gulf of Mexico were strategically important and bitterly contested territories between the French and the British. Mitchell warns the map's British supporters that France has almost complete control over the area, although he criticizes their ability to rule effectively, describing "Savage Indians" along the west coast, and inland "Nauchee" locals "extirpated" (massacred) by the French.

◀ **AROUND LAKE ONTARIO** French and English trading posts clustered around Lake Ontario, a flashpoint in the French and Indian War (1754–63), which began as Mitchell worked on his map. The coloured lines show how far French claims clashed with British ones. Mitchell also records the presence of the Native American Huron and Iroquois peoples in his name places and descriptions, explaining where they aided the British – and abetted the French.

▲ **NOVA SCOTIA** The Treaty of Utrecht enlarged Britain's control of Nova Scotia to include French territory in eastern Québec and Maine (named here as "Acadia"). Mitchell's first attempt to show the Nova Scotia coastline came under severe criticism in the form of John Green's Nova Scotia map, published in 1755. Mitchell accepted the criticism and revised his map in later editions in acknowledgment of the area's importance.

IN **CONTEXT**

Mitchell's colonial mapping was part of a history stretching back to the 17th-century British plantations in Virginia, where settlers drew lines on maps to indicate ownership, regardless of local claims. In 1783, as British diplomats annotated their copy of the Mitchell map with King George III's preferred territories, it became known as "George III's map".

▲ **Flemish engraver and editor** Theodor de Bry produced this early map of colonial Virginia showing settlers' territories.

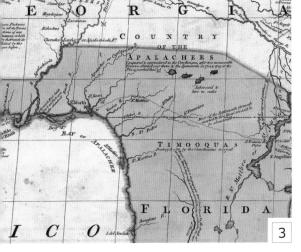

▲ **GEORGIA AND FLORIDA** Georgia and Florida's borders, including the southern Appalachians, were strenuously contested by Britain, France, and Spain. Mitchell pushes the British border further south than Spain agreed. To the west, Mitchell declares Georgia as British due to their early established "factories" (trading posts), while on the coast he seems to claim an "abandoned" Spanish fort.

▲ **ISLAND OF NEWFOUNDLAND** The French ceded much of Nova Scotia's coastline to the British in 1713–14, including the islands of St Pierre and Miquelon, anglicized here as "St Peters" and "Micklon". On the northwest coast, Mitchell tries to erase French rights to what he calls "Cape Rich" (Port au Choix).

Indian World Map

1770 ▪ TEMPERA ON CLOTH ▪ 2.6M × 2.61M (8FT 6¼IN × 8FT 6¾IN)
▪ STAATLICHE MUSEEN, BERLIN, GERMANY

SCALE

UNKNOWN

The spread of Islam into Asia, bringing with it a diverse range of Muslim, Greek, and Latin intellectual traditions, resulted in a variety of maps that intermingled cultures, languages, and geography. Nowhere can this diversity be seen more vividly than in this late 18th-century Indian world map. Although its creator is unknown, based on the exquisite vignettes covering the work it appears to have been made by a miniaturist. The map is typical of the style of painting found in the 18th century in Rajasthan in the west of India or in the Deccan plateau to the south. Its linguistic diversity also points to these regions as its source: it features descriptions written in Arabic, Persian, and Hindi (in the Devanagari script).

History, geography, and mythology

This is an extraordinarily eclectic map. It draws on a lost Islamic work by the notable 15th-century Arab sailor, Ibn Mājid, called *Secrets of the Sea*, and also acknowledges European geography with its depictions of Portuguese vessels and the Atlantic islands. Although the map remains surprisingly indebted to Ptolemy's *Geography* (see pp.24–27), it chooses a typically Islamic orientation, with south at the top of the map. Religious centres such as Mecca are shown, as are the great world cities including Constantinople (now Istanbul), and yet strange mythological creatures also roam the land and sea, in echoes of the medieval mappae mundi tradition (see pp.56–59). Its defining feature takes the form of the exploits of Alexander the Great, called "Iskandar" in Islamic traditions, and retold through the *Iskandarnamah* ("Book of Alexander"). Various aspects of his life and adventures are described on the map, representing just how common his story was across most pre-modern cultures, and how central he was to understanding geography.

IN **CONTEXT**

Alexander the Great's exploits were known to classical Greek, Roman, Persian, Islamic, and Indian cultures. In the Qur'an, "Dhul-Qarnayn" is believed to represent Alexander. By the 7th century CE, he had become mythologized in Islamic Persian literature as "Iskandar", a relative of the Persian king, Darius III. A 13th-century Persian poet, Nizāmī Ganjavī, composed a version of the *Iskandarnamah*, which influenced various other versions of the story, and was possibly also consulted by this map's maker.

▲ **Alexander holds court** in China, depicted in an 18th-century Persian *Iskandarnamah* manuscript.

> The map's **mythological content** derives largely from the *Iskandarnamah*, the Alexandrian romance, many versions of which were composed in Asia 99

SUSAN GOLE, BRITISH AUTHOR AND MAP HISTORIAN

Visual tour

KEY

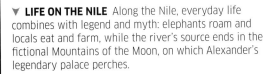

▶ **MECCA AND THE KAABA** Islam's holy sites are shown in great detail. Most of the Arabian Peninsula is taken up depicting Mecca and the surrounding mountains. At the very centre is the "Kaaba", lying at the heart of Al-Masjid Al-Haram, Islam's holiest site. Its architecture is shown with some precision.

▼ **LIFE ON THE NILE** Along the Nile, everyday life combines with legend and myth: elephants roam and locals eat and farm, while the river's source ends in the fictional Mountains of the Moon, on which Alexander's legendary palace perches.

1

2

3

▲ **THE PORTUGUESE IN INDIA** European influences can be discerned in the Indian Ocean, where a red caravel (small ship) and dinghy are anchored, identified as Portuguese by a nearby inscription. The artistry and detail here suggest the map's maker was a miniaturist, rather than a trained cartographer.

▶ **THE SPRING OF LIFE** The map's most remarkable feature is the black, rectangular "Spring of Life" in the far north, near the pole. Muslim belief attributed its discovery to the Old Testament's Moses, although its location here is strikingly reminiscent of the biblical Paradise found on medieval Christian mappae mundi.

4

5

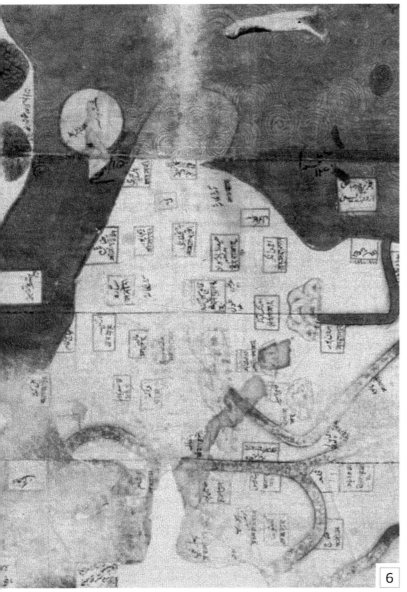

6

7

◄ **ALEXANDER'S ARMY**
In one of the most explicit depictions of the story of Alexander's confrontation with the monstrous mythical race of Gog and Magog, the map shows the emperor "with the men who asked for his help against the people of Gog and Magog, and the wall built for their defence".

◄ **INDIA** The map's Indian origins give this area topographical detail – more than 50 place names – although its shape still echoes Ptolemy's beliefs. Surprisingly, Sri Lanka is drawn twice, a remnant of age-old confusion over the mythical island of "Taprobana", in the Indian Ocean (see p.65).

▲ **IBERIA AND ATLANTIC ISLANDS**
The mapmaker's confusion over European discoveries is obvious in the Atlantic. Islands discovered by Portugal are named but misplaced, while the tear-shaped land at the top right might even represent America.

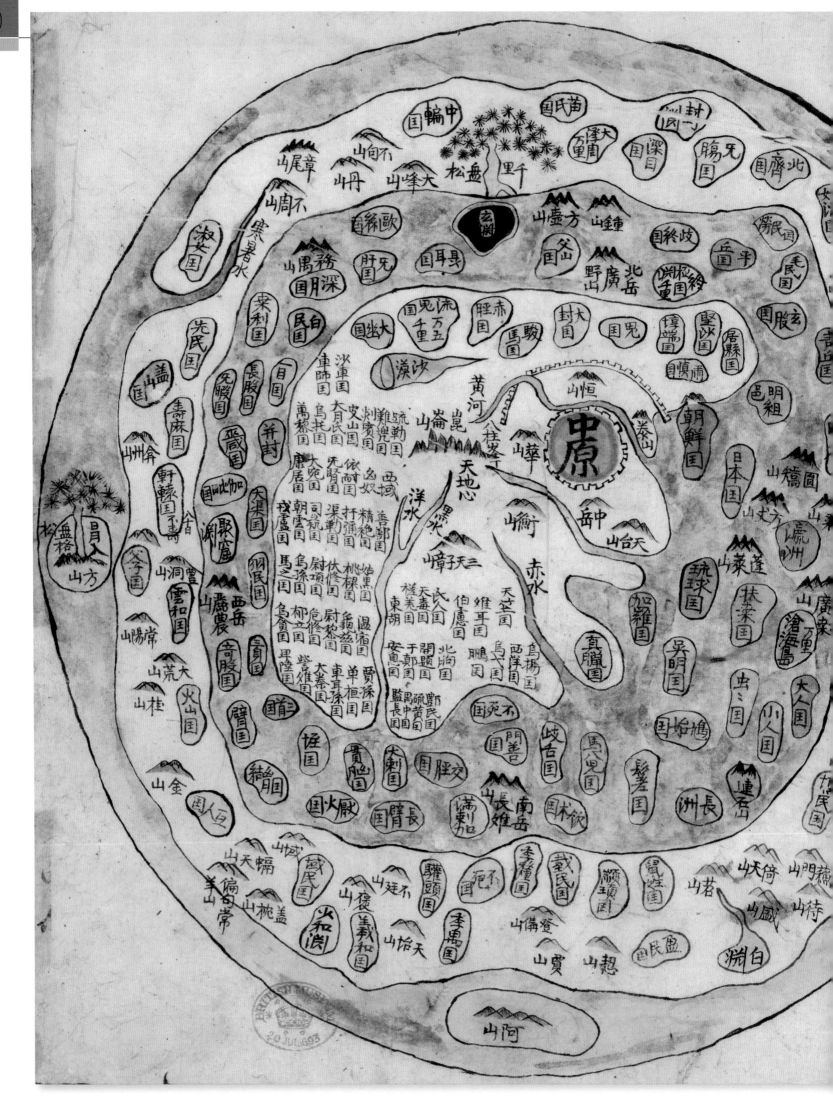

Map of All Under Heaven

c.1800 ▪ INK ON PAPER ▪ c.45CM × c.45CM (c.17¾IN × c.17¾IN)
▪ BRITISH MUSEUM, LONDON, UK

UNKNOWN

SCALE

Korea has made several important contributions to the history of cartography, but nowhere more beautifully and more mysteriously than with the style of map known as *Cho'onhado* ("Map of All Under Heaven"). The precise dates and authors of these maps remain a mystery, although the style appears to have emerged in the 16th century and reached a peak of popularity (especially among western visitors) in the late 19th century.

A view of the East

The *Cho'onhado* maps are usually oriented with north at the top, and most describe the same 143 locations. They all feature a main continent depicting *Hainei* (a term for the civilized world) and centred on China. This particular example also includes around 30 additional places, including Korea. The central continent is surrounded by an enclosed inner sea that contains over 50 place names, including Japan, Cambodia, and Siam (Thailand), which are drawn as islands. Many of the other locations are fictional, as are all of the 50 or so places, peoples, and geographical features on the outer land ring. This outer continent is in turn surrounded by a sea that encircles the world. Two trees to the east and west mark the rising of the Sun and Moon.

This map is a fascinating mix of fact and fiction, with reality giving way to fantasy as it moves outwards from the centre of the known world. Many of its names and topographical features derive from an archaic Chinese book of geographical lore, called the *Shanhai jing* ("Classic of Mountains and Seas"). Some scholars have identified Buddhist influences in its legendary trees and mountains. Of particular interest are the fantastical descriptions of people and places in the extreme east (or Pacific region) and the west (Europe) – the latter is often described on *Cho'onhado* as the "great wasteland". To the east of the map are the lands "where wood is eaten", and others inhabited by "hairy people". To the north, things become even stranger, with lands of "tangled string", "people without bowels", and a land labelled simply, "uninteresting". Although they are the product of very different beliefs to the Christian maps of the same era, the fabulous places and monsters of the *Cho'onhado* demonstrate a shared stylistic affinity with Christian mappae mundi (see pp.58–61).

> That **human nature** is the same the world over is shown by [this] **map of the world** as conceived by the Chinese and Korean mind

HOMER B. HULBERT, AMERICAN MISSIONARY AND JOURNALIST

Visual tour

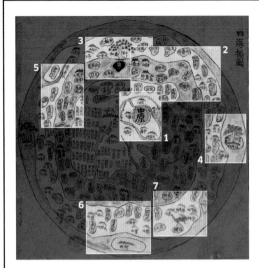

KEY

▶ **CHINA, THE YELLOW RIVER, AND THE GREAT WALL** As many of the locations on the map come from classical Chinese texts, it is no surprise that China is shown in detail. It dominates the central continent, and is labelled in red as *Zhōngguó* (the Middle Kingdom). To the north, the Great Wall is shown bisecting the Yellow River, which is coloured appropriately. A series of mountains are drawn, variously labelled as "Everlasting", "Great" and "Kunlun". This last range mixes reality with myth, as the Kunlun mountains, one of the longest mountain ranges in Asia, run across northern China, and were regarded as the home of various ancient gods.

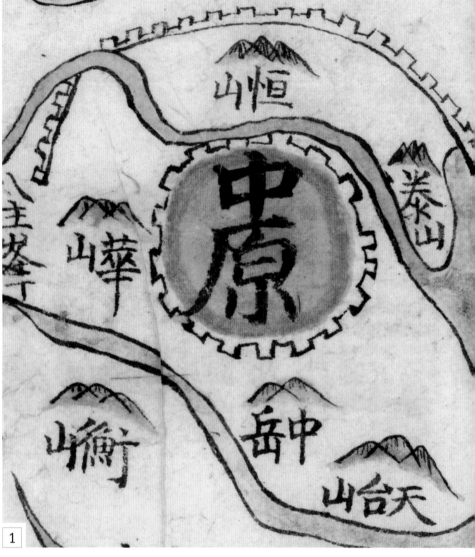

1

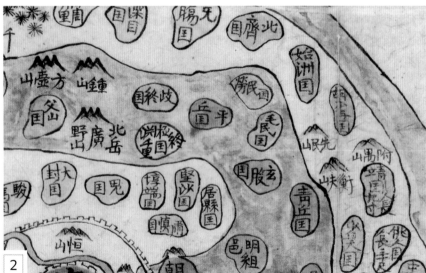

2

▲ **KOREA AND JAPAN** On the *Cho'onhado*, China is represented as the largest country, followed by Korea, then Japan. The south Asian world was defined predominantly by these three great powers. The Korean peninsula is marked by a series of locations in yellow, and to its bottom right sits Japan, shown in red. On the far right is the fantastical region of the outer ring-like continent, including the mythical lands of "Light Men" and "the Elder Dragon".

▶ **TREE AND SURROUNDING ISLANDS** In the far north, imaginary elements are combined with surprisingly realistic suggestions of the northern hemisphere. A vast mythical tree dominates, taken from the Chinese *Shanhai jing*, ("Classic of Mountains and Seas") and described as covering "a thousand li", approximately 500km (300 miles). But to its right is a "covered lake", a "land without sunshine", and "people with deep-set eyes", which could be interpreted as garbled but approximate descriptions of the Arctic region.

3

▼ THE LAND OF FUSANG To the far east of China is the mythical island of Fusang, where the sun rises, and the fabled mulberry tree of life is located. Speculation continues into the location of Fusang – many 19th century scholars even believed it to be the west coast of America. However, the map's fusion of real and sacred topography hints that it may have existed in a spiritual rather than a terrestrial space.

▶ THE WASTELAND OF EUROPE In a striking inversion of the monstrous Asian and African races portrayed on Christian mappae mundi, the islands on the *Cho'onhado* around northern Europe are represented as barbaric wastelands. These include islands of "vain", "abject", and "monkish" inhabitants, and the "Land of White People", showing a discernible, if limited, awareness of the western world.

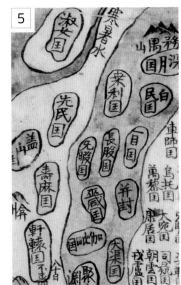

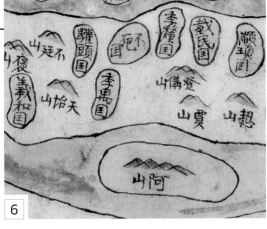

▲ ON THE MARGINS Perhaps the most fanciful of all the map's regions is described at its southernmost point – the furthest from the centre of the map, and therefore from civilized, Chinese-centred culture. It includes "Angry Mountain" to the right, and lands to the left "where it is hard to live", populated with "people with animal's heads" and other monstrosities.

◀ AMAZONS The map's southeastern areas are a series of fantastical locations. The green mountainous regions are labelled "Waiting" and "Medicine-gate". To the right the "Land of Black People" is next to the "Land of Women", compared fancifully by some to the mythical Amazonian nation of female warriors in South America.

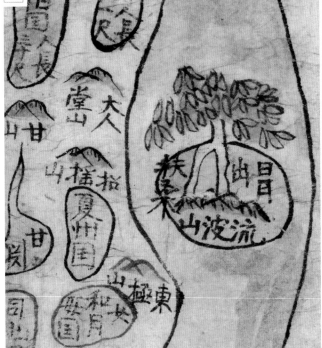

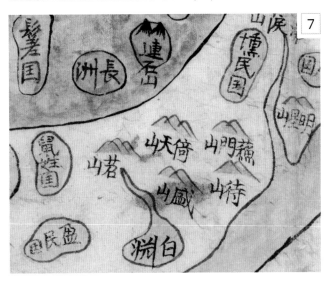

IN **CONTEXT**

The maps of most pre-modern cultures mix myth with reality, based on their particular beliefs. The *Cho'onhado* emerged from south Asian Taoist, Buddhist, and Confucian traditions based on living in harmony with nature. Korean maps were particularly influenced by *hyŏngse* (shapes and forces) – a way of placing dwellings in relation to geological features such as mountains, rivers, and forests. The Korean practice of *p'ungsu* was inherited from the Chinese tradition of feng shui, but was shaped by the rugged terrain of Korea: it saw mountains and rivers as veins and arteries, conveying energy across the land, which was seen as a body. The *Cho'onhado's* arterial rivers, prominent mountains, and trees all suggest an influence from these beliefs.

▶ This map, made around 1860, shows the Korean peninsula.

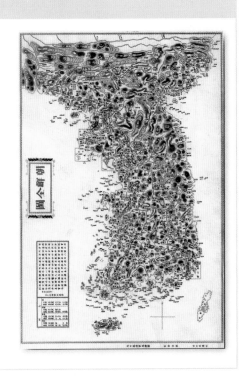

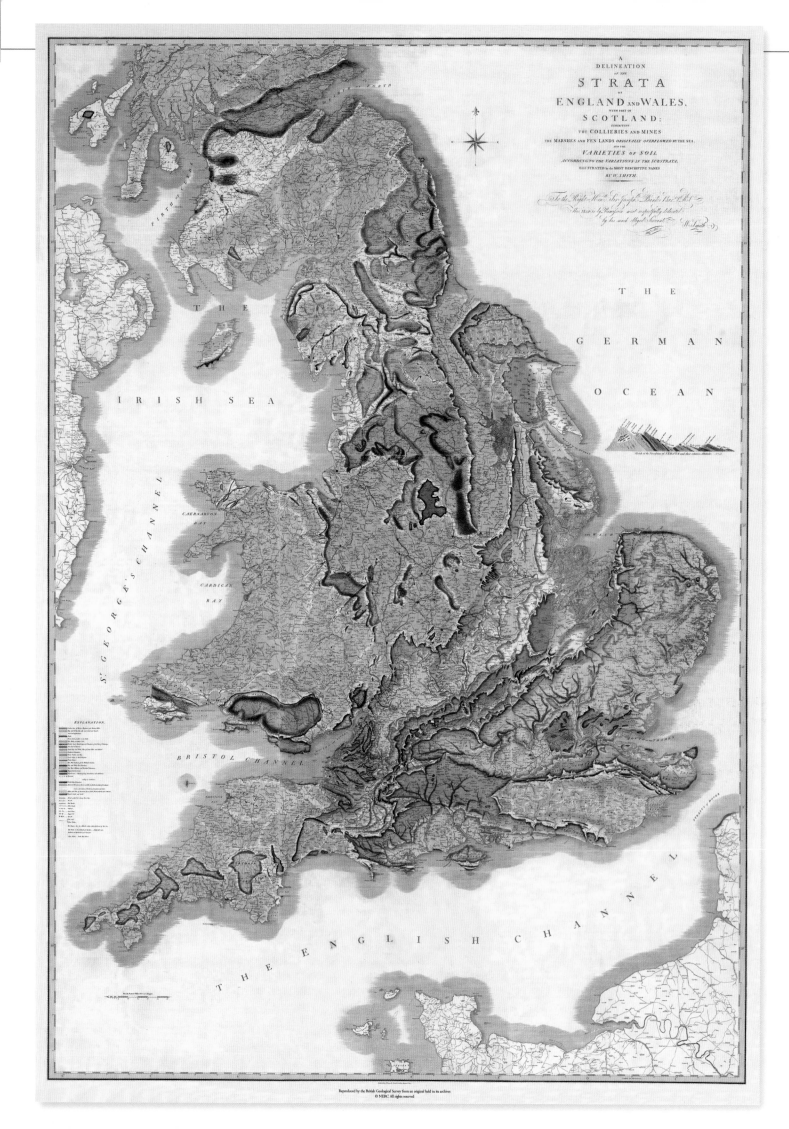

A
DELINEATION
OF THE
STRATA
OF
ENGLAND AND WALES,
WITH PART OF
SCOTLAND;
EXHIBITING
THE COLLIERIES AND MINES
THE MARSHES AND FEN LANDS ORIGINALLY OVERFLOWED BY THE SEA,
AND THE
VARIETIES OF SOIL
ACCORDING TO THE VARIATIONS IN THE SUBSTRATA,
ILLUSTRATED by the MOST DESCRIPTIVE NAMES
BY W. SMITH

A Delineation of the Strata of England and Wales with Part of Scotland

1815 ■ ENGRAVING ■ 2.4M × 1.8M (8FT × 6FT) ■ GEOLOGICAL SOCIETY, LONDON, UK

SCALE

WILLIAM SMITH

The 18th century witnessed the rise of a new form of science – geology. The composition and formation of the Earth's surface became the subject of great academic debate, but it also attracted the commercial interest of mining companies looking for both precious metals, such as gold and silver, and industrial minerals, such as iron and coal. Inevitably, maps were needed to identify the various levels of the Earth's strata (layers of rock and soil).

In response to this demand, a geologist named William Smith published this map of England in 1815. It was a completely new way of mapping the Earth, showing what the world looked like underground. Smith's vast map was made of 15 sheets at a scale of 2.5cm to 8km (1 inch to 5 miles) or 1:316,800, which Smith believed was ideal for examining different geological strata. It was made with the help of the country's foremost engraver and cartographer, John Cary, who first prepared a base map using copperplate engravings, showing the land's natural topography but without shading or relief. Thousands of geological features were then added by hand using watercolour, completed by teams of skilled craftswomen who worked to Smith's exacting requirements.

The natural order

Smith's great discovery, based on his extensive practical experience as a coal-mining surveyor, was that the different layers of rock – or strata – such as sandstone, siltstone, mudstone, and coal, occurred in a clear order that also matched their fossil contents. He called this "faunal succession", and noted that it could be used to date different kinds of rock by the fossils found within each stratum. Using this information, Smith applied distinctive, variable colour tone and shade techniques

to his map, illustrating different strata and their age in three-dimensional relief. The result is not only an extremely beautiful map but also a scientifically precise one, and the first of its kind in the world.

> It is a map whose making signifies the start of an era… marked ever since by the excitement and astonishment of scientific discoveries

SIMON WINCHESTER, *THE MAP THAT CHANGED THE WORLD*

WILLIAM **SMITH**

1769–1839

The son of a blacksmith, William Smith was a largely self-taught geologist and surveyor. However, he struggled to gain public recognition for his many achievements.

While working in coal mines in Somerset, in the southwest of England, Smith began to notice distinct geological patterns in the rocks, and set about amassing a vast collection of fossils to support his argument of "faunal succession". He began to publish geological maps in 1799, and in 1815 he published his now-famous map. Tragically, more established figures in the geological world appropriated his ideas, and he became bankrupt, spending two years in debtors' prison. It was only in the last few years of his life that Smith received recognition for his achievements, including the Geological Society's inaugural Wollastone Medal in 1831.

Visual tour

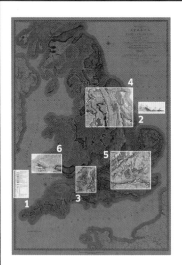

KEY

▼ **EXPLANATION** Maps usually require a legend to explain their scale, contour, or political geography; but a geological map needs to explain the structure of the layers of rock. This "Explanation" appears at the side of Smith's map. His caption places the strata in colour-coded order, reflecting the rocks' colours. These range from dark blue "London Clay", through chalk, limestone, granite, sandstone, and, of course, black coal.

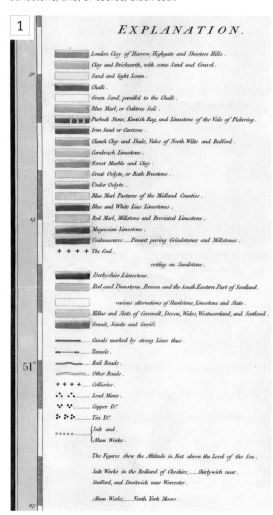

1

EXPLANATION.

London Clay of Harrow, Highgate and Shooters Hills.
Clay and Brickearth, with some Sand and Gravel.
Sand and light Loam.
Chalk.
Green Sand, parallel to the Chalk.
Blue Marl, or Oaktree Soil.
Purbeck Stone, Kentish Rag, and Limestone of the Vale of Pickering.
Iron Sand or Carstone.
Clunch Clay and Shale, Vales of North Wilts and Bedford.
Cornbrash Limestone.
Forest Marble and Clay.
Great Oolyte, or Bath Freestone.
Under Oolyte.
Blue Marl Pastures of the Midland Counties.
Blue and White Lias Limestones.
Red Marl, Millstone and Breciated Limestone.
Magnesian Limestone.
Coalmeasures.——Penant paving Grindstones and Millstones.
+ + + + The Coal.

resting on Sandstone.

Derbyshire Limestone.
Red and Dunstone, Brecon and the South Eastern Part of Scotland.

various alternations of Hardstone, Limestone and Slate.

Killas and Slate of Cornwall, Devon, Wales, Westmoreland, and Scotland.
Granit, Sienite and Gneiss.

Canals marked by strong Lines thus.
Tunnels.
Rail Roads.
Other Roads.
+ + + + Collieries.
⋮ ⋮ Lead Mines.
⋮ ⋮ Copper D.º
⋮ ⋮ Tin D.º
Salt and
○○○○○ Alum Works.

The Figures shew the Altitude in Feet above the Level of the Sea.

Salt Works in the Redland of Cheshire,——Shirlywich near.
Stafford, and Droitwich near Worcester.

Alum Works,——North York Moors.

▶ **SKETCH OF STRATA** In the blank area delineating the North Sea (labelled on the map as the "German Ocean", its name since classical times), Smith has added an extraordinary cross-section of England and Wales, slicing through the Earth from west to east and showing the country's elevation. It drops from the granite and slate of Mount Snowdon, towering at more than 1,000m (3,500ft), down through the coal and marl deposits of central England, and ends with the low-lying "Brickearth" and "London Clay" of the Thames Valley.

2

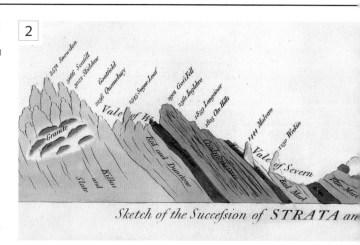

Sketch of the Succession of STRATA an

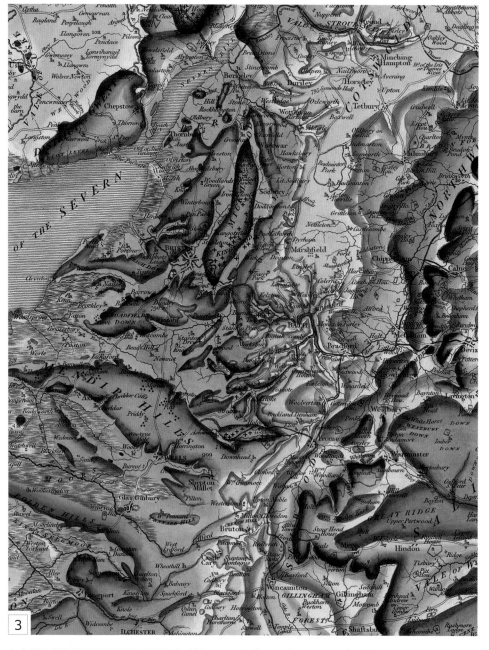

3

▲ **BATH AND SURROUNDING AREA** Smith's greatest discoveries were made when he was a surveyor for the Somerset Coal Company, working underground near the famous Georgian spa town of Bath. Here, the distinctive yellow "Bath oolite" – which Smith calls "Bath Freestone" – dominates. Dating from the Middle Jurassic period, the oolite is edged with red marl, with black coal deposits to the north.

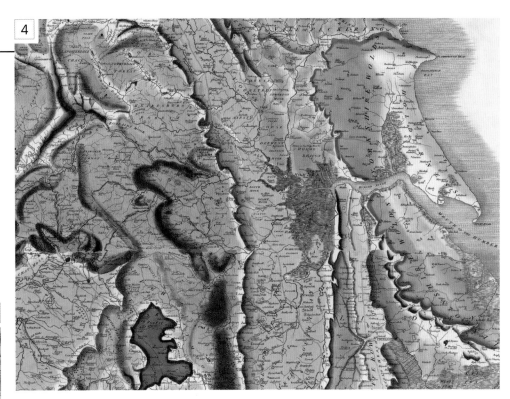

► YORKSHIRE AND THE PENNINES In 1794, Smith travelled to Yorkshire with members of the Somerset Coal Canal to inspect the region's mining potential. His map highlights the rich black coal reserves shown running through the lowland areas of South Yorkshire. The bleak carboniferous limestone of the Pennines is to the northwest, running into the Peak District's "Derbyshire Limestone".

◄ LONDON The geology of London is starkly captured in Smith's colours. It sits in a basin composed of the blue clay and ochre chalk, dating back to the Palaeozoic era, with the Thames snaking in from the sea to the east. The chalky green Chilterns lie to the north, with the North Downs to the south of the city.

ON **TECHNIQUE**

William Smith led the field of geological mapmaking, but he was not the first to depict geological features in graphic form. Georges Cuvier and Alexandre Brongniart published what they called a "geognostic" map of the Paris Basin in 1808. Smith, however, was the first to develop techniques to show both geological time and space on a map. He utilized copperplate engraving at a suitable scale to illustrate strata using contour lines, as they appeared just on or below the Earth's surface. His clever use of shading also allowed him to represent depth, or the sedimentation of time on his maps. This influenced the later evolutionary theories of Charles Darwin by showing how stratified deposits of fossils developed in complexity over long periods of time.

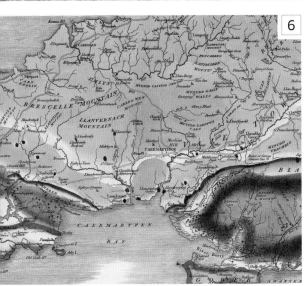

◄ SOUTHWEST WALES There is no more dramatic example of how Smith's geological mapmaking would affect the face of the country's physical landscape than his depiction of southeast Wales. A great swathe of low-lying black coal deposits are shown running from Swansea in the west to Cardiff in the east, some reaching a thickness of over 1,800m (6,000ft). They are surrounded by the higher, red-coloured Brecon sandstone to the north.

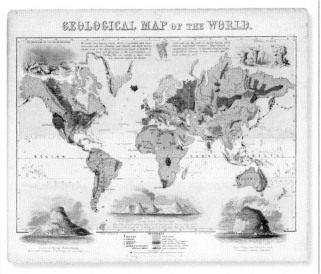

▲ **This geological map of the world,** engraved by John Emslie and published in 1850, follows the style of William Smith's ground-breaking map.

Japan, Hokkaido to Kyushu

1821 ■ RICE PAPER ■ 86CM × 1.04M (2FT 9¾IN × 3FT 5IN) ■ LIBRARY OF CONGRESS, WASHINGTON, DC, USA

SCALE

INŌ TADATAKA

Mapping the coastline of Japan is a daunting task. Although the main islands of Hokkaido, Honshu, Shikoku, and Kyushu account for 97 per cent of the country, the remaining three per cent is spread out over 6,848 smaller islands. However, amateur cartographer Inō Tadataka was so determined to map the whole of Japan that he devoted

INŌ **TADATAKA**

1745-1818

Surveyor and cartographer Inō Tadataka' played a vital part in mapping modern Japan. Born in a coastal village, he was adopted by a wealthy family in the town of Sawara at the age of 17.

There, he ran the family sake-brewing and rice-trading business, until he retired at the age of 49 to study geography, astronomy, and mathematics. In 1800, he began an official survey of Japan, producing a series of maps. In addition, Inō wrote several scholarly works about surveying and mathematics.

the final 16 years of his life, and a good portion of his own money, to surveying the whole country. In total, he travelled nearly 35,000km (21,750 miles), producing 214 large-scale maps, three of which are shown here. Although Inō died before he could complete his survey, it was finished by his team in 1821.

Unlike previous Japanese cartographers, Inō used modern traverse surveying techniques, and prioritized mathematical and geographical accuracy over political or historical considerations. However, the maps were so detailed, showing all major roads, rivers, and coastlines, that the Tokugawa shogunate (military rulers), worried about foreign invasion, kept them largely hidden from the public. It was only when imperial rule was restored to Japan in 1868 that the maps were more widely distributed. Until well into the 20th century, Inō's were the definitive maps of Japan, and the data gathered during his surveys informed many subsequent maps.

▲ **HOKKAIDO** Called Ezochi on Inō's map, this island was renamed Hokkaido in 1869.

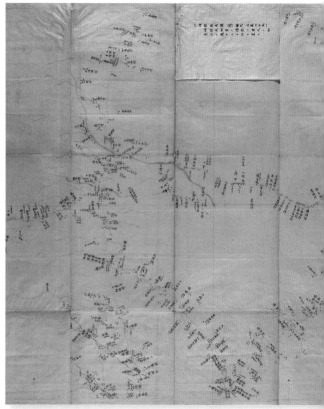

▲ **HONSHU, EDO** In 1821, Kyoto was the official capital of Japan, but Edo (now Tokyo) was the political centre.

Visual tour

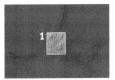

KEY

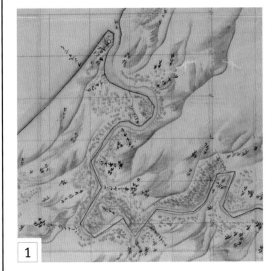

1

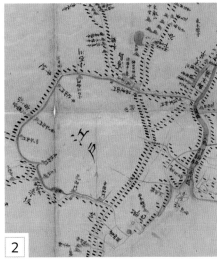

2

3

▲ **ISHIKARI PLAIN** Inō meticulously details the path of the Ishikari River as it meanders into the Sea of Japan on the western side of Hokkaido. The area to the south of this river is the site of modern-day Sapporo, officially founded in 1868 and now the largest city on Hokkaido.

▲ **EDO CASTLE** Here, Inō depicts the Tokugawa shogunate's seat of power, Edo Castle, and its sprawling, heavily fortified grounds. After the demise of the shogunate, the castle became the Imperial Palace in the renamed city of Tokyo.

▲ **HAMANAKO** Lake Hamana, also known as Hamanako, is the 10th largest lake in Japan and feeds directly into the Pacific Ocean. On his map, Inō includes copious notes relating to the settlements around the lake, as well as details of the topography.

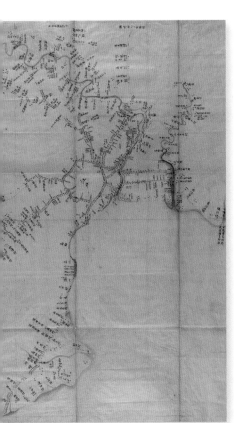

▲ **HONSHU, SHIZUOKA PREFECTURE** Sand dunes and a large lake characterize the southern coastline of Honshu.

"Indian Territory" Map

1839 ▪ PAPER ▪ 1.26M × 1.61M (3FT 5IN × 4FT 4½IN) ▪ NATIONAL ARCHIVES, WASHINGTON, DC, US

SCALE

HENRY SCHENCK TANNER

Before the American Revolutionary War (1774–83), European settlers had largely been confined to 13 colonies on the Atlantic (east) coast. However, with the formation of the United States, the colonists began to expand their interests to the fertile lands to the west. In many cases, this meant "removing" (relocating) the indigenous tribes, so in 1839 the Bureau of Topographical Engineers, led by Captain Hood, began collecting information about the Native American territories. Draughtsman Joseph Goldsborough Bruff then overlaid the data on an existing map of the east

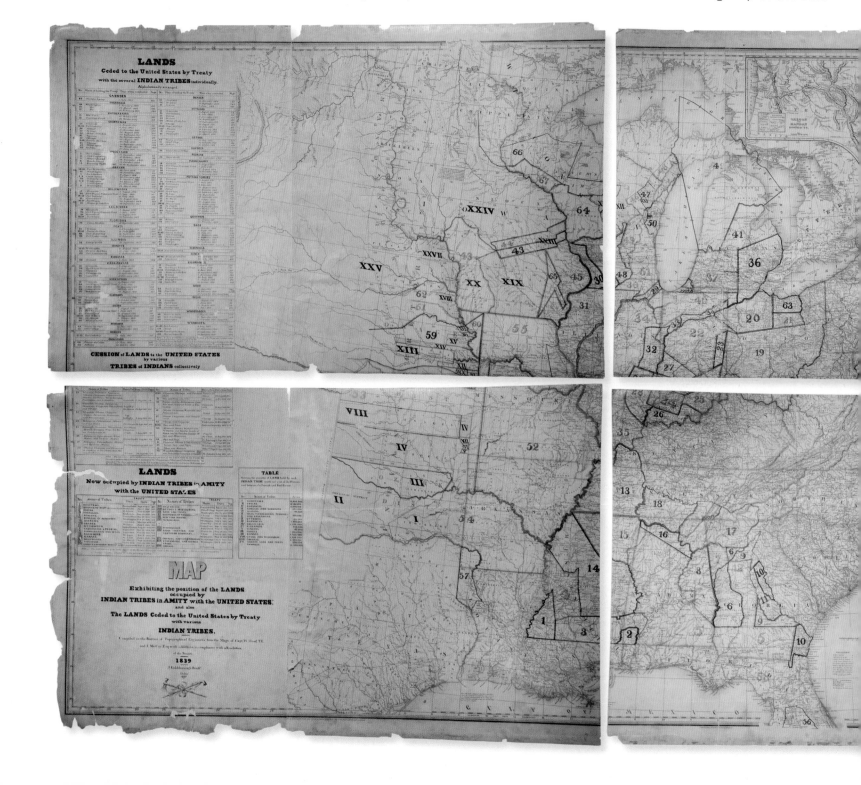

and central areas of the United States, which had originally been created by Henry Schenck Tanner. Coloured zones relate to the lands Native Americans had already ceded in treaties and their new lands to the west – which became known as "Indian Territory" – and tables around the edges contain supplementary information.

While other maps of "Indian Territory" were also made during this period, Tanner's edited map was the most complete, and was the only one created on such a large scale. There are only three surviving copies; this one, found in the National Archives in Washington, DC, is believed to be the version consulted by the US Congress as they planned the various stages of Native American removal.

HENRY **SCHENCK TANNER**

1786-1858

Born in New York but predominantly based in Philadelphia, Henry Schenck Tanner was an eminent cartographer and map publisher during a golden age of American mapmaking.

Tanner devoted his career to the art of mapmaking, both as a cartographer and a publisher. His most famous works included the *New American Atlas*, published in 1823; the first ever map of Texas in 1830, before it even became an official state; and *A Geographical and Statistical Account of the Epidemic Cholera from its Commencement in India to its Entrance into the United States*, published in 1832 in response to the 1817 global cholera epidemic. In 1846, Tanner also released the popular *New Universal Atlas* of the world and created many notable travellers' guides, state maps, wall maps, and pocket maps.

Visual tour

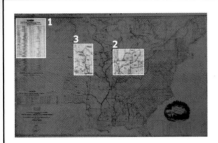

KEY

▼ **EXTRA INFORMATION** The two tables on the top left of the map relate to land that the Native Americans have ceded, while the bottom two detail the new territories, using numbers and Roman numerals, which also appear on the main map. In this concise table, Bruff details how much land, by acre, each tribe has been given in the relocation programme.

▷ **INDIAN TERRITORY** The Native American tribes were pushed further west, in the direction of what would later become the state of Oklahoma. The Roman numerals reflect the new territories and the numbers the old lands, so some more fortunate tribes, such as the Otoes and Missourias (61 and XVIII) did not have far to go. Others, of course, were far less fortunate.

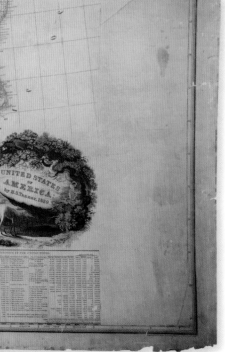

▲ **ORIGINAL SETTLEMENTS** The coloured numbers relate to the tables on the left of the map. This region, south of Lake Huron and Lake Michigan was originally home to many different tribes. The area marked "19", for example, was inhabited by Wyandets, Delawares, Shawnees, Ottawas, Chippewas, and at least seven other tribes. However, in the relocation programme, many tribes were separated from their former neighbours.

John Snow's Cholera Map

1854 ▪ PAPER ▪ 30CM × 28.5CM (1FT × 11¼IN) ▪ BRITISH LIBRARY, LONDON, UK

JOHN SNOW

SCALE

This map of London's Soho district represents the birth of modern epidemiology – the medical study of the patterns, distribution, and control of diseases in a population. It was created by John Snow, a physician who rejected the popular belief that diseases such as cholera were caused by a "miasma" (bad air), but was unable to provide an alternative theory. In 1854, while investigating a cholera outbreak in Soho that killed over 500 people in just 10 days, Snow plotted the incidence of deaths on a street map of the area. The dramatic results showed a pattern of

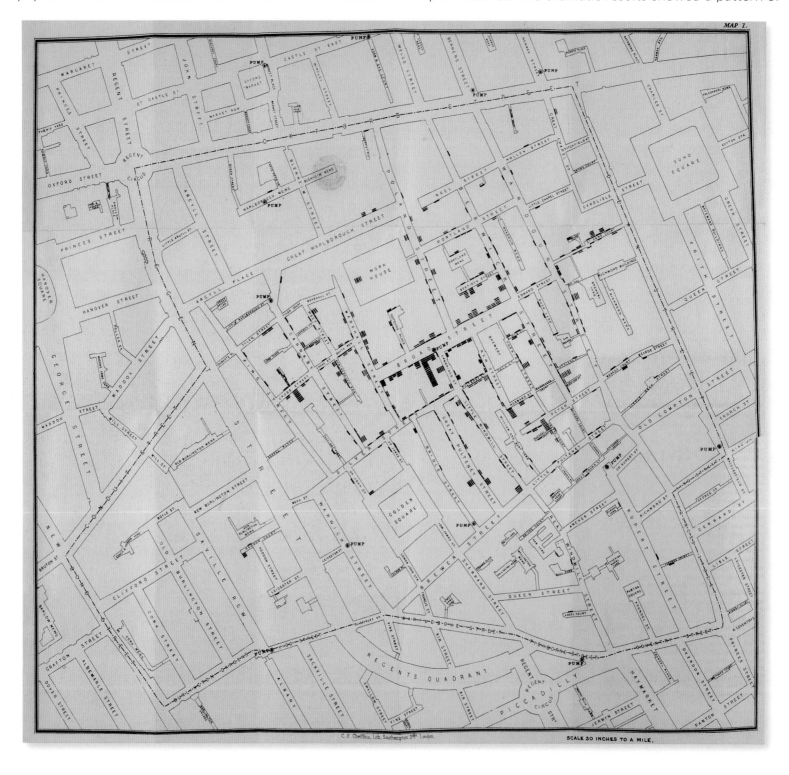

SCALE 30 INCHES TO A MILE.

deaths clustered around one particular public water pump on Broad Street, which was used by residents for drinking water. Snow concluded the pump's water was infected with faeces from a nearby cesspit. He used the map to persuade authorities to close the pump, and new cholera cases immediately declined (the local authorities, uncomfortable with his theory of the faecal transmission of cholera, later restored the Broad Street pump to working order).

The map changed the way we visualize complex data, and enabled Snow to explain the spread of disease – it also showed how statistics could be used in the form of a map to affect public policy. Snow's methods still underpin modern epidemiology and preventative healthcare.

JOHN **SNOW**

1813–1858

Widely regarded as one of the founders of modern epidemiology, John Snow was a British physician who rose to prominence from humble origins.

Born in York in the north of England, he trained in London before pioneering the use of surgical anaesthesia, including administering chloroform to Queen Victoria during childbirth, which was extremely unusual at the time. In 1850, he founded the Epidemiological Society of London, which anticipated his subsequent work on London's cholera outbreaks. Although the initial impact of Snow's work was short-lived due to the prevailing theories of the time, in the long-term he changed the face of modern medicine. Snow continued his struggle for health reform until his sudden death from a stroke in 1858.

Visual tour

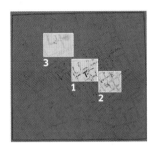

KEY

▶ **BROAD STREET** By projecting each cholera case as a black bar onto a Soho street map, Snow could immediately discern the origin and spread of disease: the Broad Street water pump was obviously the source of the infection.

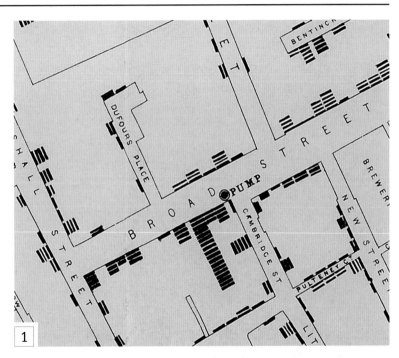

1

ON **TECHNIQUE**

Snow supported his findings with ground-breaking use of "control groups". This involved studying the health of a group of residents from the same part of Broad Street, but who worked at a local brewery with its own pump-well. This group remained healthy, indicating that the problem was with the Broad Street pump. Snow's methods anticipated the advent of "germ theory", and identified the spatial distribution and transmission of a disease, even if its original cause remained unknown. This insight would prove vital in subsequent battles with diseases such as malaria and HIV.

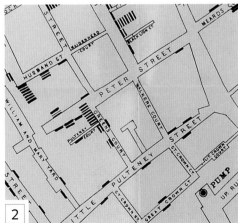

2

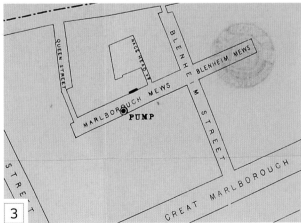

3

▲ **PETER STREET** Southeast of Broad Street lay Peter Street, an area that suffered from a significant but relatively equal distribution of infection. Snow could ascribe this to the increasing distance from Broad Street's pump. Moving further south, cases of cholera tailed off almost completely.

▲ **MARLBOROUGH MEWS** Northwest of Broad Street, in the more affluent area around Marlborough Mews, Snow recorded just one cholera case, despite the proximity of a public water pump. This was because different pumps drew on different water sources – some were more liable to infection than others.

▲ **The infected pump** on Broad Street.

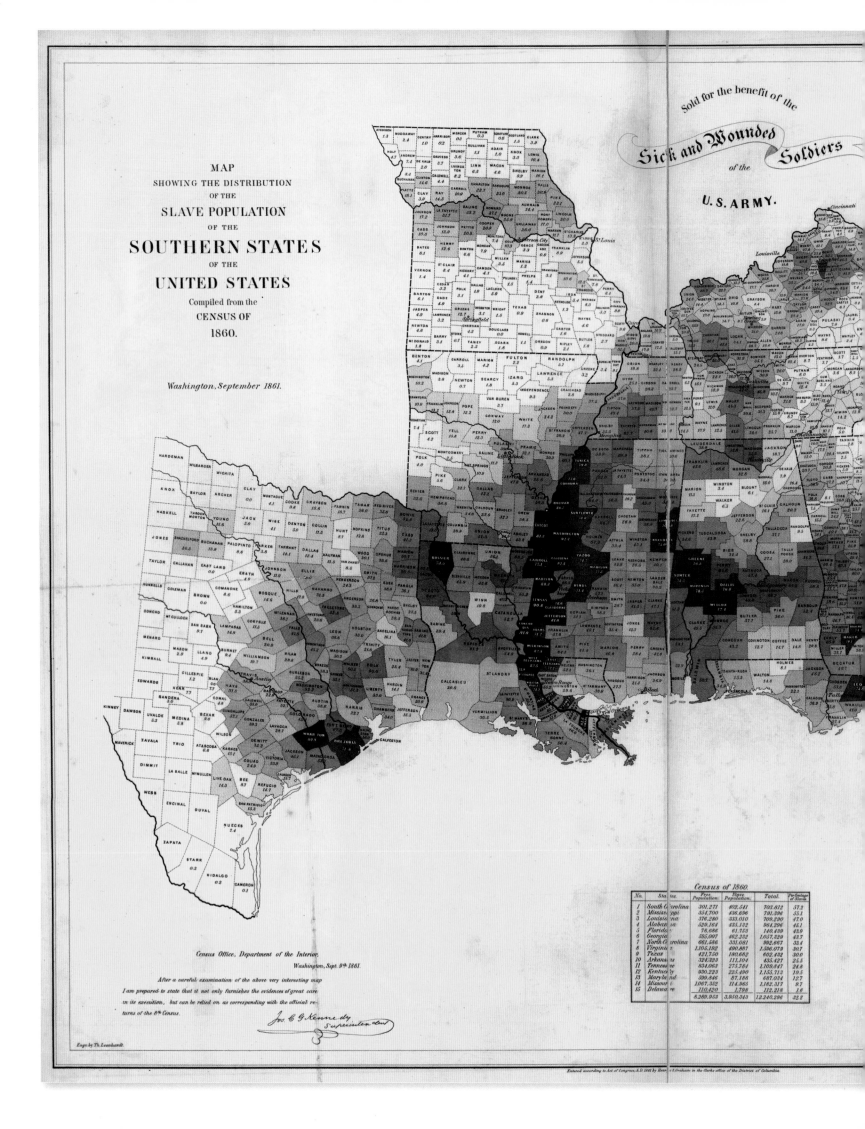

MAP
SHOWING THE DISTRIBUTION
OF THE
SLAVE POPULATION
OF THE
SOUTHERN STATES
OF THE
UNITED STATES
Compiled from the
CENSUS OF
1860.

Washington, September 1861.

Sold for the benefit of the

Sick and Wounded Soldiers

of the

U. S. ARMY.

Census Office, Department of the Interior.
Washington, Sept. 9th 1861.

After a careful examination of the above very interesting map
I am prepared to state that it not only furnishes the evidences of great care
in its execution, but can be relied on as corresponding with the official re-
turns of the 8th Census.

Jos. C. G. Kennedy
Superintendent

Engr. by Th. Leonhardt.

No.	States.	Free Population.	Slave Population.	Total.	Percentage of Slaves.
1	South Carolina	301,271	402,541	703,812	57.2
2	Mississippi	354,700	436,696	791,396	55.1
3	Louisiana	376,280	333,010	709,290	47.0
4	Alabama	529,164	435,132	964,296	45.1
5	Florida	78,686	61,753	140,439	43.0
6	Georgia	595,097	462,232	1,057,329	43.7
7	North Carolina	661,586	331,081	992,667	33.4
8	Virginia	1,105,192	490,887	1,596,079	30.7
9	Texas	421,750	180,682	602,432	30.0
10	Arkansas	324,323	111,104	435,427	25.5
11	Tennessee	834,063	275,784	1,109,847	24.8
12	Kentucky	930,223	225,490	1,155,713	19.5
13	Maryland	599,846	87,188	687,034	12.7
14	Missouri	1,067,352	114,965	1,182,317	9.7
15	Delaware	110,420	1,798	112,218	1.6
		8,289,953	3,950,343	12,240,296	32.2

Census of 1860.

Entered according to Act of Congress, A.D. 1861 by Henry S. Graham in the Clerks office of the District of Columbia.

Slave Population of the Southern States of the US

1861 ▪ LITHOGRAPH ▪ 66CM × 84CM (2FT 2IN × 2FT 9IN)
▪ LIBRARY OF CONGRESS, WASHINGTON, DC, USA

SCALE

EDWIN HERGESHEIMER

In 1860, a national census of the US estimated the population at more than 31 million. Of this, nearly 4 million people were slaves, compared with only 700,000 in 1790. It was a shocking statistic, but the issue of slavery divided the nation: in 1860 abolitionist Abraham Lincoln won the presidential election, but before he was even inaugurated, seven southern, pro-slavery states had declared their secession from the northern states. These southern states formed the Confederate States of America in 1861 and the country soon descended into a four-year civil war. Six months into the American Civil War, the US Coast Survey's cartographer Edwin Hergesheimer, an abolitionist, published this landmark map.

New techniques and social change

Hergesheimer used the new chloropleth mapping techniques, in which geographical areas were shaded or patterned according to the distribution of statistical data on a specific subject – in this case, the percentage of the population who were registered slaves. The map explained the secession crisis as being driven by the economic fear of abolishing slavery, and was designed to be "sold for the benefit of the sick and wounded soldiers" fighting against the Confederate states. President Lincoln even had his portrait painted consulting the map, which he used to follow the movement of his Union troops through states with high slave populations.

EDWIN **HERGESHEIMER**

1835-1889

An immigrant from Germany, Hergesheimer was a political radical who had been forced to flee his native country after the failed revolutions of 1848. Like many people disappointed with the lack of political reforms in Europe, he ended up in America.

On his arrival, Hergesheimer joined the United States Coast Survey, one of the Federal Government's agencies, which had been founded in 1807. He was employed in the Survey's Drawing Division as a cartographer, working under its superintendent, Alexander Bache. Along with the commercial engraver Henry Graham, the three men collaborated together to produce the famous document that Abraham Lincoln called his "slave map". Hergesheimer went on to make topographic maps of key battles and fortifications in the Civil War.

NOTE.

It should be observed, that several counties appear comparatively light. This arises from the preponderance of whites and free blacks in the large towns in those counties, such as ——— Henrico Co. Va., Norfolk Co. do., Shelby Co. Tenn., Davidson Co. do., St Louis Co. Mo., Orleans Co. La., Charleston Co. S.C. &c.

The figures in each County represent the percentage of slaves viz.: Amherst Co. Va. 46½ are slaves in every 100 inhabitants; Wayne Co. N.Carolina 38½ are slaves in every 100 inhabitants &c.&c.

Scale of Shade.

	Less than 10 per cent
	10 & less than 20 per ct.
	20 . . . 30 . .
	30 . . . 40 . .
	40 . . . 50 . .
	50 . . . 60 . .
	60 . . . 70 . .
	70 . . . 80 . .
	80 per ct. & upwards.

Drawn by E. Hergesheimer.

Visual tour

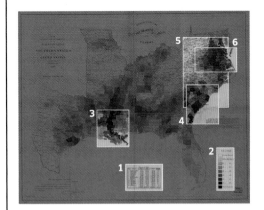

KEY

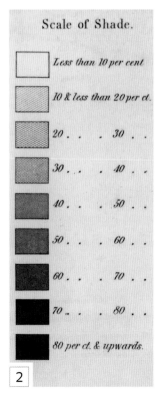

Census of 1860.

No.	States.	Free Population.	Slave Population.	Total.	Per Centage of Slaves.
1	South Carolina	301,271	402,541	703,812	57.2
2	Mississippi	354,700	436,696	791,396	55.1
3	Louisiana	376,280	333,010	709,290	47.0
4	Alabama	529,164	435,132	964,296	45.1
5	Florida	78,686	61,753	140,439	43.9
6	Georgia	595,097	462,232	1,057,329	43.7
7	North Carolina	661,586	331,081	992,667	33.4
8	Virginia	1,105,192	490,887	1,596,079	30.7
9	Texas	421,750	180,682	602,432	30.0
10	Arkansas	324,323	111,104	435,427	25.5
11	Tennessee	834,063	275,784	1,109,847	24.8
12	Kentucky	930,223	225,490	1,155,713	19.5
13	Maryland	599,846	87,188	687,034	12.7
14	Missouri	1,067,352	114,965	1,182,317	9.7
15	Delaware	110,420	1,798	112,218	1.6
		8,289,953	3,950,343	12,240,296	32.2

1

▲ **CENSUS OF 1860** To emphasize his objectivity and his scientific and statistical credentials, Hergesheimer included a table based on the 1860 census, showing the free and enslaved populations of the states on his map. It showed huge variations in numbers and percentages: at 57 per cent, South Carolina had the larger percentage of slaves, but Virginia had the largest number at nearly half a million. An average of over 30 per cent of the southern states' population were slaves.

▶ **SCALE OF SHADE** Hergesheimer's challenge was to transform the census data into a graphic, and project it across the southern states. His "Scale of Shade" reveals his decision to break down the data into the percentage of slaves rather than total populations. He also chose to depict these percentages within counties, rather than just states, producing enormous variations in distribution. Shading from lower percentages in lighter tones to higher percentages in darker tones was a particularly striking innovation.

Scale of Shade.

	Less than 10 per cent
	10 & less than 20 per ct.
	20 . . . 30 . .
	30 . . . 40 . .
	40 . . . 50 . .
	50 . . . 60 . .
	60 . . . 70 . .
	70 . . . 80 . .
	80 per ct. & upwards.

2

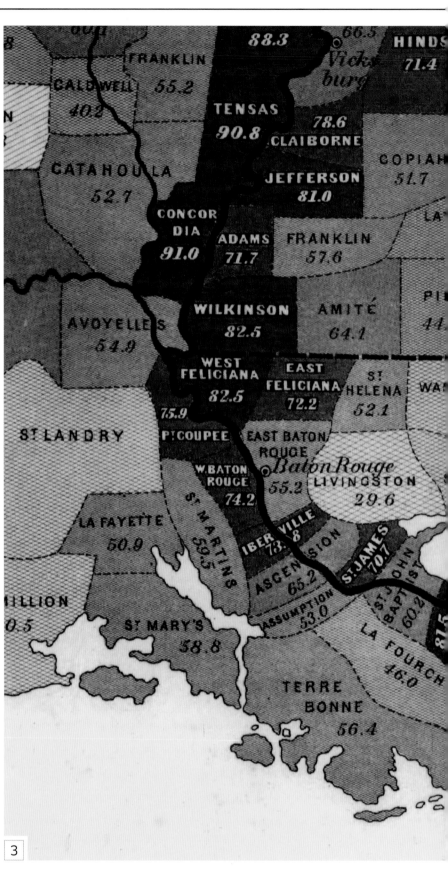

3

▲ **MISSISSIPPI RIVER** Snaking through the counties within Mississippi and Louisiana, the river crosses some of the areas with the highest rates of slavery in the South. These were part of the Civil War's Western Theatre, stretching from east of the Mississippi. Plotting a route through the densest areas of slavery almost exactly reproduces the Union army's March to the Sea, led by Major General William Sherman in 1864, a crucial military turning point that struck at the economic heart of Confederate slavery.

▼ **GEORGIA** Hergesheimer showed concentrations of slavery in areas where the plantations farming tobacco, sugar and, in Georgia's case, cotton, were most profitable. Even in Georgia, a state with 43 per cent of its population registered as slaves, percentages fluctuated from one county to another, depending on economic conditions. Writing of Georgia the campaign, Union General Sherman promised "the utter destruction of its roads, houses, and people will cripple their military resources. I can make the march and make Georgia howl".

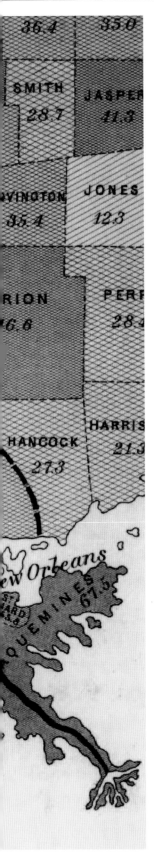

It should atively light. T free blacks in Henrico Co Va.

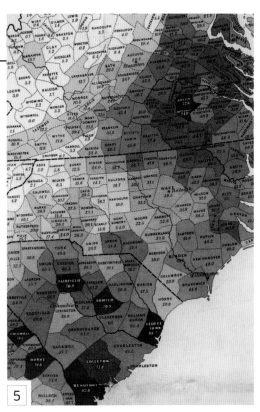

▲ **THE APPALACHIANS** Huge differences in slave populations are recorded across the Appalachian Mountains, a region deeply divided between Union and Confederate sympathies, and one that saw 80 counties ravaged by civil conflict, with some towns changing sides repeatedly. Many mountainous communities had developed their own agricultural methods without recourse to slavery, and they resented supporting the slave-owning planter class in the lower-lying areas.

IN **CONTEXT**

The 19th century was the great age of thematic mapping, showing the geographical nature and spatial distribution of various subjects or themes, previously invisible on maps, such as crime, disease, poverty, or race. It emerged out of the growth of statistical methods and national censuses (first held in England and France in 1801), and led to a range of thematic maps, such as this one by André-Michel Guerry, and Charles Booth's poverty map (see pp.206–209).

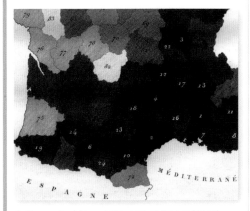

▲ **Guerry's chloropleth map** of France (1833) shows the frequency of donations to the poor per number of residents.

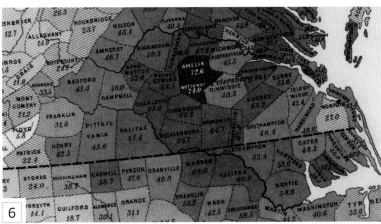

▲ **VIRGINIA** The state of Virginia holds a special place in American constitutional history. It is known as the "Mother of Presidents", and many of the nation's founding fathers were born there, including Thomas Jefferson, James Madison, and the first president, George Washington; it was also home to the oldest legal general assembly in the Americas. As Hergesheimer revealed, however, it was also a key member of the Confederacy that embraced slavery, with counties such as Amelia registering slave populations of over 70 per cent.

Dr Livingstone's Map of Africa

1873 ■ PRINT ON LINEN ■ 72CM × 79CM (2FT 4¼IN × 2FT 7IN)
■ ROYAL GEOGRAPHICAL SOCIETY, LONDON, UK

SCALE

DAVID LIVINGSTONE

Towards the end of his last expedition to Africa, the great Scottish explorer David Livingstone drew this map of central Africa. It was based on the surveys, drawings, and observations he made during his final years in Africa, from 1866 until his untimely death from malaria and dysentery in Zambia in May 1873.

The map traces in red his last expedition from Zanzibar on the coast, through modern-day Tanzania, Zambia, and the Congo, in search of a solution to the centuries-old debate as to the source of the River Nile. Livingstone believed that the river's source lay in Lake Tanganyika, around which much of his journey circulated, as can be seen on the map. Although Livingstone was wrong – most geographers now accept that the Nile's source emanates from Lake Victoria to the north – his last expedition discovered a wealth of new lakes and rivers, and brought international attention to the Arab slave trade in the area. His map reveals the African interior for the first time, and frees it from the centuries of fantastic myths and monstrous legends perpetuated by so many earlier maps.

> I go back to Africa to try to make an open path for commerce and Christianity

DAVID LIVINGSTONE

DAVID **LIVINGSTONE**

1813-1873

Born into a Scottish Protestant family, David Livingstone worked in cotton mills as a child before training in medicine and joining the London Missionary Society.

His African missionary work began with several epic journeys, during the course of which he became the first European to see the Victoria Falls and travel up the Zambezi River. He carried out his expeditions motivated by the belief that commerce and religion could civilize Africa and eradicate slavery, and effectively "vanished" from the Western world for several years. In 1871, his fellow explorer Henry Morton Stanley found him in Tanzania and uttered the immortal (but probably apocryphal) line, "Dr Livingstone, I presume?"

Visual tour

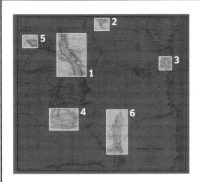

KEY

▶ **LAKE TANGANYIKA (AND UJIJI)**
Livingstone's route shows he had already traversed the lake's shores in 1868, before returning in 1871. By Ujiji on the eastern shore, the map records "Mr Stanley's arrival, Oct. 28 1871", one of the most famous meetings in the history of exploration (see p.199). "Scenery lovely", Livingstone noted as they both explored northwards.

▶ **LAKE VICTORIA**
At the top of the map, Livingstone drew Lake Victoria, or "Nyanza" according to local Bantu languages. The first European to discover it was John Hanning Speke in 1858, who, contrary to Livingstone's beliefs, claimed it was the Nile's source.

▲ **ZANZIBAR** In January 1866, Livingstone returned to Africa for the last time. He landed at Zanzibar off the Tanzanian coast, a British protectorate and the centre of the Arab slave trade. Livingstone hated its squalor, calling it "Stinkibar", and in March he left for the mainland with his 36-strong team.

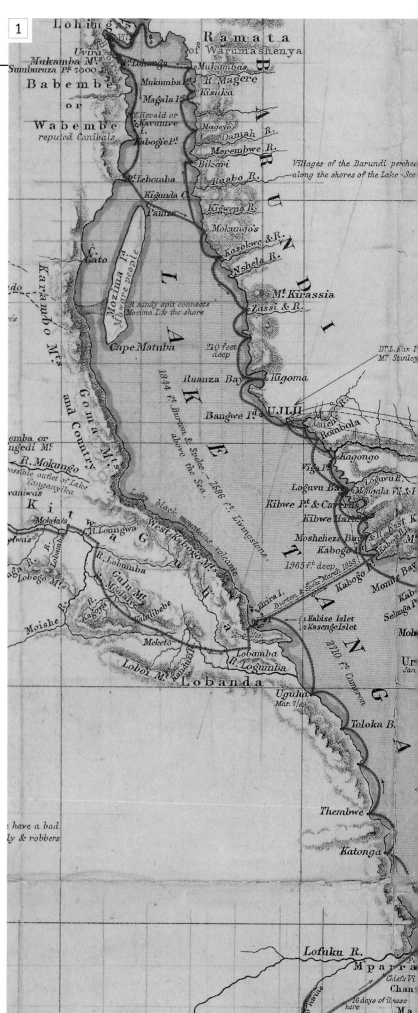

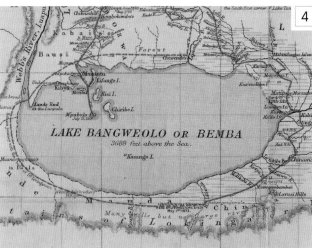

◀ **LAKE BANGWEULU** By July 1868, with few supplies left and his health failing, Livingstone crossed the flooded swamps of the Upper Congo River Basin in search of the Nile's sources. On 18 July, he became the first European to sight Lake Bangweulu, in what is now Zambia.

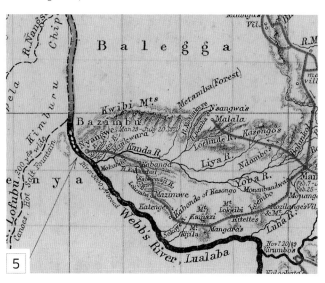

▲ **NYANGWE** In March 1871, Livingstone reached Nyangwe on the Lualaba River. What the map does not record is that four months later he witnessed the horrific massacre of 400 Africans, (mainly women) by Arab slave traders from Zanzibar, in an attempt to intimidate the locals. Livingstone's published account of the massacre caused international outrage, and is widely credited with precipitating the suppression of slavery in Zanzibar.

IN **CONTEXT**

Livingstone's exploration of Africa preceded the "scramble for Africa", during which European imperial powers carved up the continent according to their colonial interests. In Joseph Conrad's 1899 novel *Heart of Darkness*, the protagonist looks at a map of Africa "marked with all the colours of the rainbow", indicating French (blue), German (purple), and British (red) possessions. However, the reality was less well-defined. For example, by 1900, more than 15.5 million sq km (6 million sq miles) of territory claimed by the British in Africa had yet to be mapped.

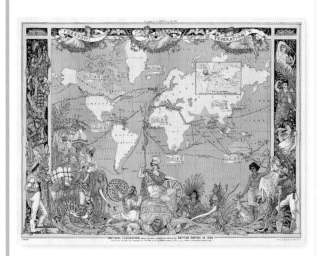

▲ **This world map from 1886** shows the territories of the British empire in pink; the "Scramble for Africa" had only just begun.

▲ **LAKE NYASA** Livingstone had discovered Lake Nyasa on a previous expedition in 1859. Returning in 1866, he rounded its southern end then headed northwest towards Lake Tanganyika. However, his original team deserted him, and by this stage only three or four of them remained.

Missionary Map

LATE 19TH CENTURY ▪ METALLIC PAINT AND PENCIL ON MUSLIN ▪ 1.27M × 1.75M (4FT 2IN × 5FT 9IN)
▪ SMITHSONIAN AMERICAN ART MUSEUM, WASHINGTON, DC, USA

UNKNOWN

SCALE

Angels and demons stalk the Earth in this apocalyptic missionary map, designed to recruit believers to 19th-century preacher William Miller's Adventist cause. After studying the Bible, Miller concluded that there would be a Second Coming (or Advent) of Christ, and even predicted a date – 22 October 1844. He mass-produced an educational chart outlining his beliefs (see right), which influenced other works, including this map.

The map includes the entire African continent, suggesting that it could have been used by late 19th-century missionaries working in Africa. The large figure on the left is based on the dream of King Nebuchadnezzar from the Book of Daniel, with a head of gold, breast and arms of silver, belly of brass, legs of iron, and feet of clay, each representing an ancient civilization. Elsewhere, figures from the Bible adorn the landscape. The map is not entirely faithful to Miller's vision, however: as his prediction of the Second Coming proved inaccurate, it omits all his numbers and calculations.

IN **CONTEXT**

William Miller is widely seen as the Adventist movement's founder. He fervently believed in the Second Coming of Christ, the final battle between good and evil, and a fiery end to Earth. Miller, along with one of his followers, Joshua V. Himes, a Boston minister and early US adopter of the printing press, produced the *Miller Chart* in 1842.

▶ The *Miller Chart* laid the foundations for contemporary Adventists. The large figure on the left also appears on the missionary map.

Visual tour

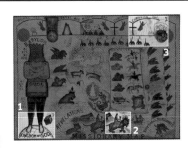

KEY

▶ **SCARLET WOMAN** Next to the figure is an apocalyptic vision of beasts and a woman dressed in scarlet. Here, the dragon represents Satan and the woman is supposed to be the Whore of Babylon, who represents earthly pleasures and sin.

`2`

▽ **THE DESTRUCTION OF MANKIND** The figure inspired by King Nebuchadnezzar's dream is divided into various ancient civilizations that have risen and fallen, from Babylon to Rome, with each one seen as inferior to the one that preceded it. This ends with the stone being hurled, which will ultimately destroy all mankind, and Earth will become the Kingdom of God.

`1`

`3`

▲ **THE SECOND COMING** The band across the top of the map represents the five ages of the world, from the primal age in the beginning to a final divine age. The flaming disc featuring the number "2" depicts the Second Coming of Christ, heralding the transition to the divine age.

Descriptive Map of London Poverty, 1898-9

1898-99 ■ PRINT ■ 1.6M × 2.23M (5FT 3IN × 7FT 3¼IN)
■ LONDON SCHOOL OF ECONOMICS, LONDON, UK

SCALE

CHARLES BOOTH

In the 1880s, the British press claimed that up to 25 per cent of London's population was living in poverty. Philanthropist Charles Booth was so shocked that he launched his own survey of living and working conditions in the city, employing teams of researchers to interview local people. His findings put the true figure at 35 per cent, and Booth used his data to create a "poverty map" of the capital. Published in 1889, the *Descriptive Map of London Poverty* revealed the complex social geography of 19th-century London for the first time. However, the map only showed the East End, so in 1891 Booth expanded it west to Kensington, east to Poplar, north to Kentish Town, and south to Stockwell. By 1898, Booth felt that a revision was needed, using new data, and covering an even greater area. Originally 12 separate maps, the 1898–99 map has since been stitched together digitally.

BOOTH'S CLASSIFICATION OF POVERTY

Black: Lowest class. Vicious semi-criminal.

Dark blue: Very poor, casual. Chronic want.

Light blue: Poor. 18–21 shillings a week for a moderate family.

Purple: Mixed. Some comfortable, others poor.

Pink: Fairly comfortable. Good ordinary earnings.

Red: Middle class. Well-to-do.

Yellow: Upper-middle and Upper classes. Wealthy.

A combination of colours – such as dark blue or black, or pink and red – indicates that the street contains a fair proportion of each of the classes represented by the respective colours.

CHARLES **BOOTH**

1840-1916

A businessman and philanthropist, Charles Booth famously documented the life of working-class people in London and pioneered social mapping.

Born in Liverpool, Charles Booth tried unsuccessfully to pursue a political career before turning his attention to mapping working-class poverty in London. He published an enormously influential study, *Life and Labour of the People* (1899), introduced the idea of a "poverty line", and supported the introduction of old age pensions.

Visual tour

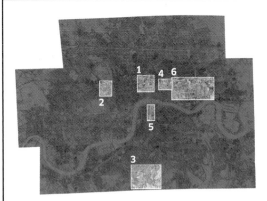

KEY

▶ **MIXED UP IN HOLBORN** One of the most striking revelations on Booth's maps was the extraordinary proximity of wealth and poverty in London. Central areas such as Holborn and Clerkenwell feature "well to do" and "fairly comfortable" streets like Hatton Garden adjacent to "vicious, semi-criminal" areas like Brooke Street.

1

2

3

◀ **MONEYED MARYLEBONE**
Some areas were almost exclusively wealthy. Marylebone, a fashionable aristocratic enclave just west of the City of Westminster, contains only small pockets of blue poverty.

▲ **PROSPEROUS BRIXTON**
The map also offers a fascinating insight into urban historical change. Today, Brixton in south London is a densely populated inner-city area, but on Booth's map it is a prosperous suburb.

▼ **POVERTY AND WEALTH IN THE HEART OF THE CITY** Booth's social mapping offers an insight into the complex historical topography of London. East of the centre, Finsbury Square, a "well-to-do" residential area, is surrounded by poor districts. Much of central London appears "poor" (pale blue), due to the concentration of low-paid manufacturing industries in this area. A similar map in the 21st century would present a very different picture.

▶ **THOROUGHFARES AND SIDE STREETS** Booth's topography of urban wealth and poverty offered a new way of understanding social development. Here the railway lines separate the "well-to-do" bourgeoisie lining the busy Blackfriars Road in Southwark from "semi-criminal" areas such as Pocock Street, to the east.

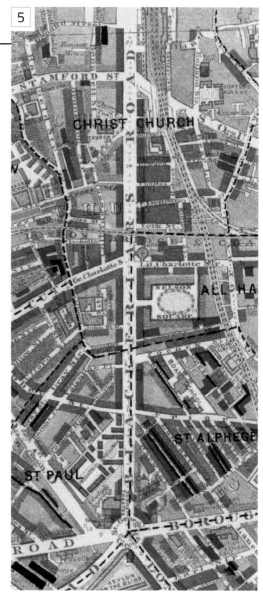

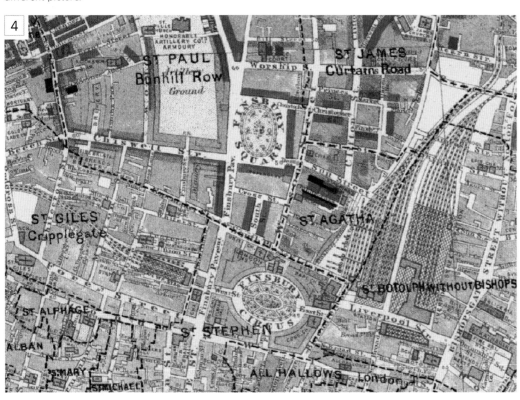

▲ **POVERTY IN THE EAST END** Traditionally, Whitechapel in London's East End was regarded as a place of poverty, criminality, and disease. Although Booth's map shows pockets of black and dark blue in places such as Stepney, the picture varies between districts. The Commercial Road area is completely red, making it the province of the middle classes. This suggests that the divisions between poverty and prosperity were less straightforward than many of Booth's contemporaries believed.

ON **TECHNIQUE**

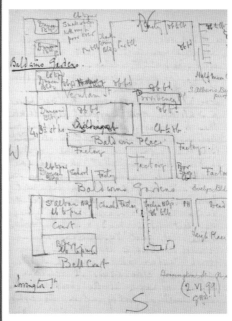

▲ **This notebook entry details** the area around Baldwin Gardens in Holborn, central London, which includes several factories.

Booth's methodology was exhaustive and complex. He asked questions about three main areas of working-class Londoners' lives: their workplace and its conditions; their homes and their urban environment; and their religious observance. Booth employed dozens of assistants to visit families, workers, factory owners, trade unionists, clergymen, and their congregations, and to interview them all in their homes or work places. Booth's assistants conducted their surveys street by street, often accompanying London School Board workers and local police on their rounds in order to gain access to as many Londoners as possible. The surveyors filled hundreds of notebooks with data, which was used to compile the statistical evidence at the heart of Booth's poverty maps.

Marshall Islands Stick Chart

DISCOVERED 1899 ▪ COCONUT LEAVES ▪ 78CM × 35CM (31IN × 13¾IN) ▪ MUSEUM FÜR VÖLKERKUNDE, BERLIN, GERMANY

UNKNOWN

SCALE

To navigate across the vast Pacific, sailors from the Marshall Islands in Micronesia, southeast of the Philippines, developed a highly original approach to seaborne mapping. Their "stick charts" were based on ocean swells and the specific nature of the Marshall Islands, 32 atolls that caused ripples over distances of around 30km (20 miles). Skilled navigators were able to detect changes in oceanic swells, indicating their distance from particular islands.

This stick chart, known as a *rebbelib*, shows the Ralik chain of islands and the intersections of the major swells running throughout them. Curved sticks indicate swells, chevrons show swells refracted around certain islands, and horizontal sticks measure distances between islands. The maps were primarily educational (consulted before a journey, but never at sea) and disappeared with the decline of inter-island canoe travel in the Marshalls in the 1950s.

IN CONTEXT

Having consulted relevant stick charts prior to departure, a Micronesian pilot would assess the wind, season, and stars, before taking a position in their canoe that enabled them to sense swells and their intersections – how they affected the pitch and roll of the canoe – and navigate accordingly. The charts were first identified by German colonists in the late 19th century, although their usage undoubtedly stretches back much further.

▲ **A modern model of a stick chart** showing oceanic swells, with shells representing atolls.

Visual tour

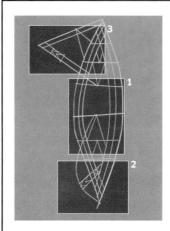

KEY

▶ **AILINGLAPLAP** The three curved sticks on the outer side represent *rilib* (eastern) and *kaelib* (western) swells. The *bōt* (central node) marks intersecting swells refracting around an atoll; here it is the Ailinglaplap coral atoll.

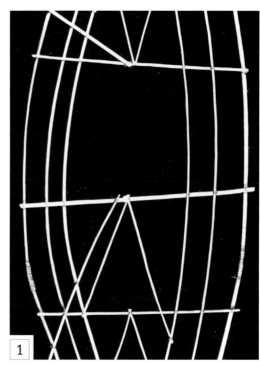

1

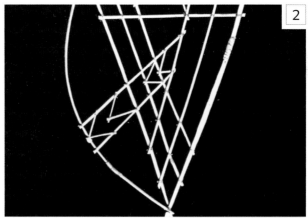

2

◀ **EBON** At the bottom of the map, several swells are disrupted by islands, in this case Namorik to the left and at the bottom, Ebon – the southernmost atoll of the entire Marshall Islands network.

▶ **WOTHO TO UJAE** The tear-shaped network defines all but two of the Marshall Islands. It extends itself westwards through two swells to connect Wotho (at the apex of the top right chevron) with Ujae (on the far left).

3

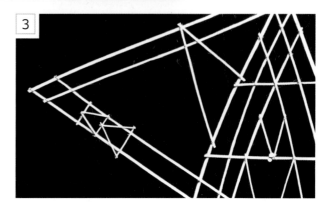

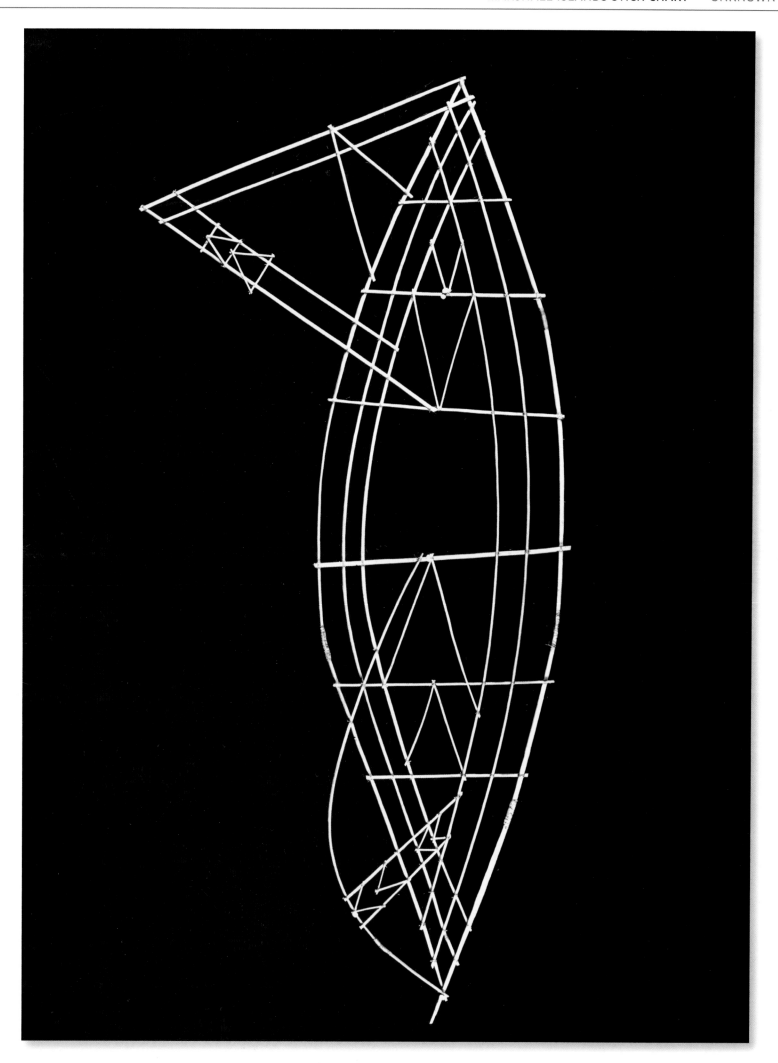

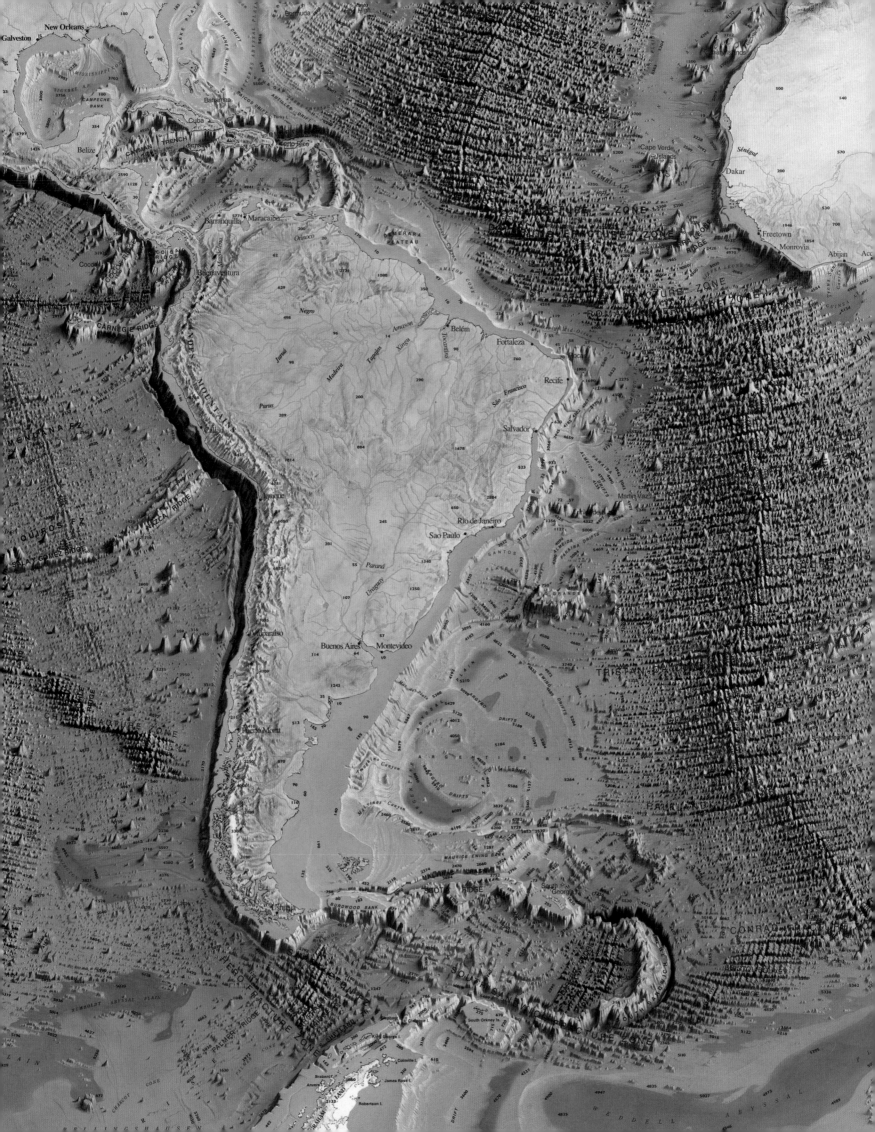

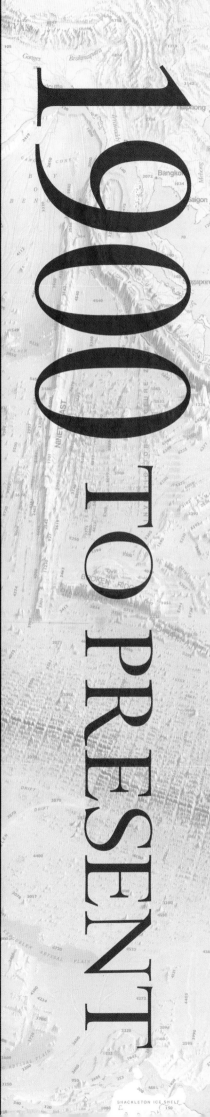

MODERN MAPPING

- International Map of the World

- London Underground Map

- Dymaxion Map

- Lunar Landings Map

- Equal Area World Map

- World Ocean Floor

- Mappa

- Cartogram

- Nova Utopia

- Google Earth

1900 TO PRESENT

International Map of the World

1909 ▪ PRINT ▪ 56CM × 1.01M (1FT 10IN × 3FT 4IN) ▪ BOSTON PUBLIC LIBRARY, BOSTON, MASSACHUSETTS, USA

SCALE

ALBRECHT PENCK

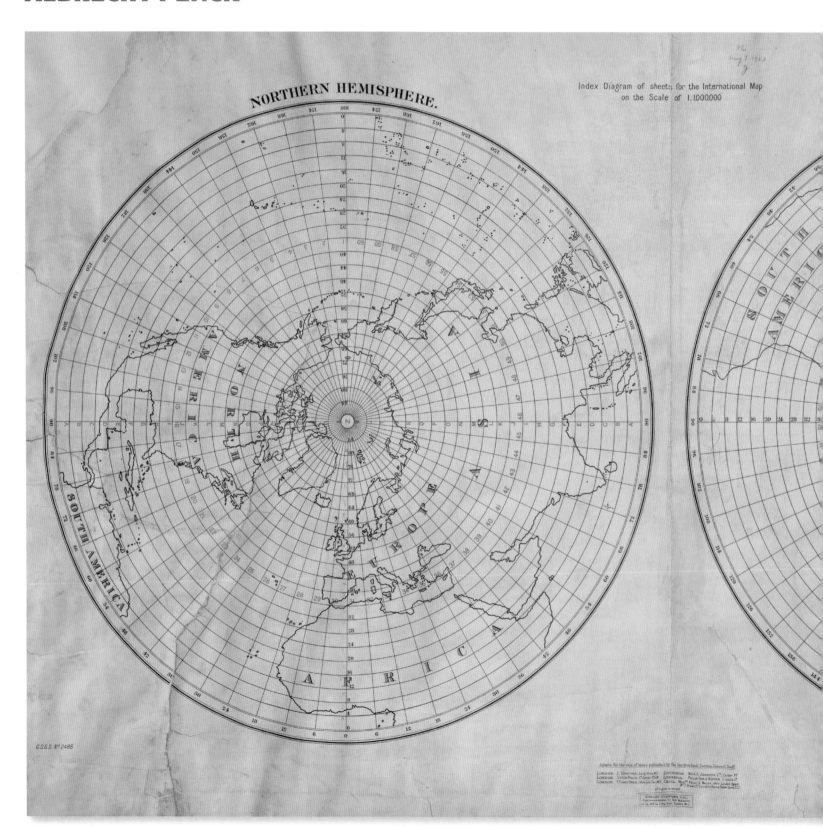

In 1891, Albrecht Penck proposed a new, standard map of the world. He argued that, in the field of world maps, there was no consistency in their scale, projection, or symbols. However, he felt that sufficient information existed for the international community to be able to work together to produce a uniform world map. Penck recommended a scale of 1:1,000,000 (1cm to every 10km or 1in for every 15¾ miles) and called it the *International Map of the World* (IMW), although it also become known as the *Millionth Map of the World*, due to its scale. This ambitious project would involve the creation of 2,500 maps covering the entirety of the Earth's surface. Each map would cover four degrees of latitude and six degrees of longitude, all with the same projection (known as a modified conic) and identical naming conventions, symbols, and even colours.

Mapping the world

Penck hoped that Europe's geographical societies and governments would cover most of the cost of the project, with the rest raised by selling copies of the maps, priced at two shillings (16 cents) each. He used the International Geographical Congresses to drive through his plans, although the spirit of global cooperation that underpinned his idea was soon overshadowed, perhaps unsurprisingly, by international and imperial rivalries. Each country was supposed to create their own maps, which presented problems for smaller nations who had neither the resources nor the expertise to survey their countries. In 1909, Penck produced this preliminary map to ascertain the extent of the project, which, judging by the numbering across its surface, was vast. Just four years later, only six maps of Europe had been completed, and shortly afterwards the US government chose to create its own 1:1,000,000 maps instead.

Nevertheless, Penck's project limped on through two world wars, suffering many setbacks, such as bombings of the archives and the production of maps that did not adhere to Penck's conventions. Penck died in 1945 and the newly formed United Nations took over the project in 1953. It was officially terminated in 1989, with less than a thousand maps finished, of which most were obsolete anyway. It was an ignoble end to a noble idea, and further proof that a map can never be truly finished.

ALBRECHT **PENCK**

1858–1945

Albrecht Penck was a distinguished geographer and geologist, whose work spanned geomorphology, climatology, and regional ecology, as well as political geography.

Penck was part of a golden age of German research into Earth sciences in the late 19th century. Born in Leipzig, he was professor at the universities of Vienna (1885–1906) and Berlin (1906–26). He is well known for his geological study of ice age stratigraphy, researching the glacial formations in the Bavarian Alps. Penck's idealistic faith in the *International Map of the World* waned after Germany's defeat in World War I. Instead, he began to develop a more exclusive belief in German nationalism and embraced German geographer Friedrich Ratzel's concept of *Lebensraum*, or living space.

Visual tour

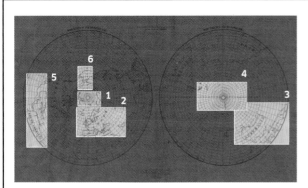

KEY

▶ **NORTH POLE** Penck
shrewdly adopted a
twin-hemispherical
projection centred
on the poles to sidestep
the issue of centring the
map on any one location
(the prime meridian
had only recently been
agreed as running
through the British
Empire's choice
of Greenwich).

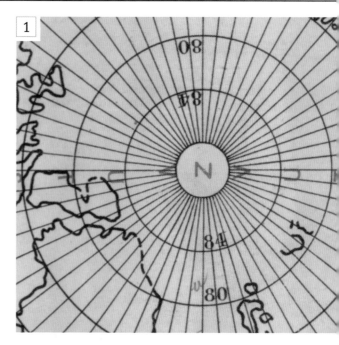

▼ **EUROPE** It is particularly striking
that Europe is shown without any political
borders. Penck seems to be making the point
that mapmaking can, and should, transcend
national boundaries.

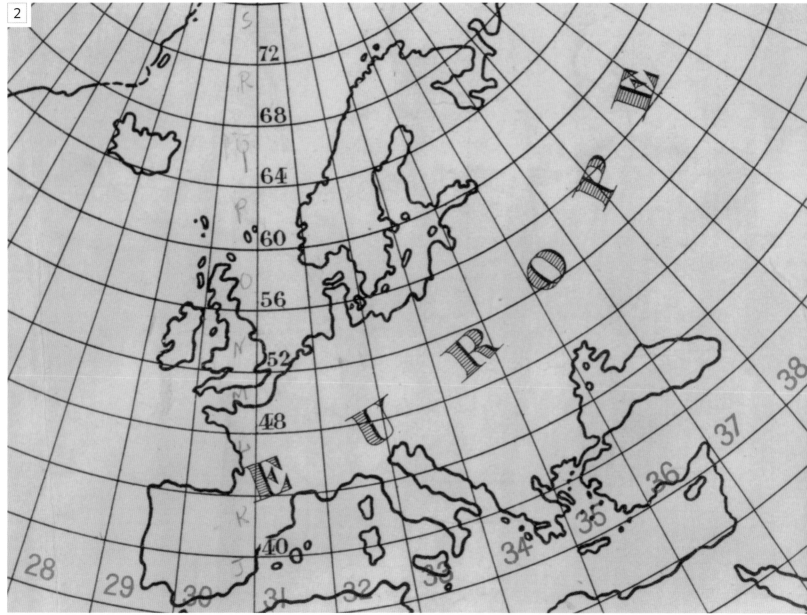

▼ AUSTRALIA AND THE INDONESIAN ARCHIPELAGO

The map's orientation still leaves certain landmasses looking marginalized. These include Australia and Indonesia, which are shown on the edge of the southern hemisphere, with almost no connection to any other continent.

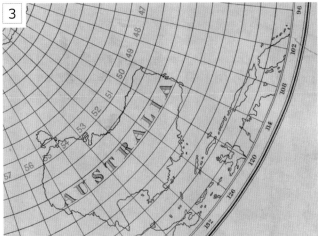

▼ ANTARCTICA

The southern hemisphere, with its projection centred on Antarctica, is particularly striking. It shows no tangible connections between the different landmasses (Africa, South America, and Australia), underlining the reasons why maps never centre themselves on this region – it has no geopolitical power.

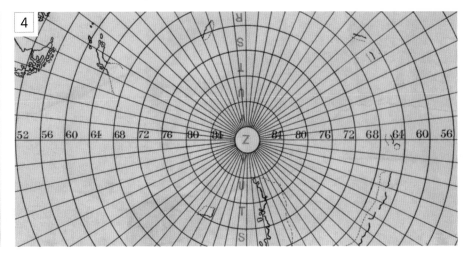

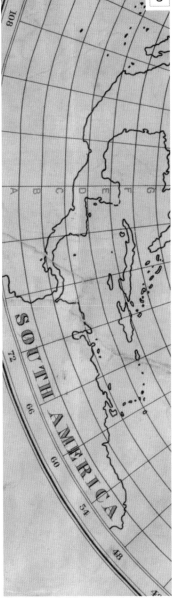

◄ SOUTH AMERICA AND THE MEXICAN GULF

Places such as South America were enthusiastic about making 1:1,000,000 maps. Its mapmakers had no interest in the European squabbles over imperial rights, they were more interested in simply getting themselves on the map. Thus, in direct contrast to North America, South America was reasonably well mapped in Penck's project.

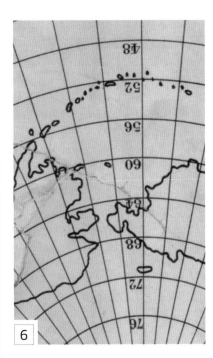

▲ THE BERING STRAIT

The orientation of Penck's map showed the closeness of the USA and Russia, separated only by the narrow Bering Strait. The 20th century's superpowers might have looked at Penck's map before developing the Cold War geographical propaganda that dominated the second half of the century.

IN **CONTEXT**

Although the IMW was built on a spirit of international cooperation, it was quickly appropriated for less high-minded activities. In 1914, the British Ordnance Survey (OS) and Royal Geographical Society (RGS) proposed to take over the IMW. The initiative was funded by MO4, part of Britain's intelligence organization with responsibilities for gathering military maps for the war effort. Throughout the conflict, the OS and RGS produced a series of 1:1,000,000 maps of Europe, the Middle East, and North Africa to help the Allied war effort. By 1939, maps on the IMW's scale were not regarded as sufficiently useful for military operations, and Allied funding for the project dried up. This was yet another area in which the IMW eventually failed.

▲ **This 1915 RGS map index** shows maps of Europe made to a scale of 1:1,000,000 – pre-World War I (outlined in red), made during the war (dark pink), and planned or in progress (light pink).

London Underground Map

1933 ▪ PAPER ▪ 15.4CM × 22.7CM (6IN × 9IN) ▪ LONDON TRANSPORT MUSEUM, LONDON, UK

SCALE

HARRY BECK

There is surely no more iconic transport map in the world than Harry Beck's plan of the London Underground train network. Used by countless millions of bewildered visitors to London since its release in 1933, it is both a great map and an archetypal piece of modern design – although the story of its creation is a testament to the problems many unique innovations face when they challenge entrenched beliefs.

Beck designed the map in response to the expansion of the London Underground network in the early 20th century, following its modest beginnings in 1863. Confronted with an overall network of more than 200 underground and overland stations covering nearly 400km (250 miles) of track, London's private rail companies had produced a mass of muddled and competing maps that caused confusion among London's commuters and visitors.

No distance at all

A draughtsman who had worked on the Underground's electrical signals, Beck understood that rail travellers were only interested in getting from one point to another in the shortest time: space was irrelevant. Passengers also required a simple diagram that enabled them to navigate the network, regardless of its relation to the ground above. Beck's solution was to design a map that made no attempt to convey distance, and that looked more like an electrical circuit board than a city's railway lines. It effectively created the idea of a transport network. Curved lines were banished in favour of vertical, horizontal, and diagonal lines, which connected stations regardless of topographical reality. He combined a modern typeface with a simplified colour code for each line, with diamonds indicating interchanges. He reduced the length of outlying lines and expanded distances between the crowded central areas, again regardless of real distance.

Beck's design was initially rejected as too radical and revolutionary in severing the link between the map and the territory it showed. It was finally accepted in 1933, although Beck's innovation was diminished throughout his life, and his genius only appreciated in recent years. Its success and enduring appeal proves that a map can sacrifice reality and still be truly great.

HARRY **BECK**

1902-1974

Born in East London, Henry "Harry" Charles Beck worked in the 1920s as a temporary engineering draughtsman at the London Underground Signals Office, and was laid off work in 1931.

He used his spare time to begin working on a new map of the rail network, based on the current design. Although it had not been commissioned by London Transport, the map was accepted in 1933, and Beck continued to improve it throughout his life, completing his last version in 1959. Unfortunately, London Transport attributed later versions of the map to other designers, much to Beck's dismay. He also proposed an improved map of the Paris Métro in the 1930s which was never used, and created a map of the rail system surrounding London.

It has touched **so many people**. The tube diagram is one of the **greatest pieces of graphic design** produced, **instantly recognizable** and **copied across the world**

KEN GARLAND, BRITISH GRAPHIC DESIGNER

Visual tour

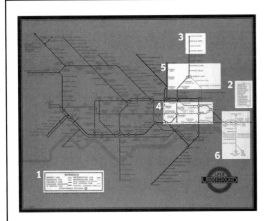

KEY

▲ **REFERENCE CHART** The key to the map acts like an old map's cartouche, allowing users to orient themselves quickly. It was given a modern look with Edward Johnston's famous eponymous sans serif typeface, which was also used for the Underground's iconic logo. The map is dated by its reference to now defunct lines such as the "Edgware, Highgate, and Morden Line" which later became today's Northern Line.

2

▲ **THE CITY AND BEYOND** The Underground was constantly evolving and connected to a vast overground rail network – much of it mapped by Beck. In the East End, just above Bow Road station, he acknowledges the growing network by indicating overland stations running northeast through Bromley and Dagenham, and ending nearly an hour away from London, in Southend on the Essex coast.

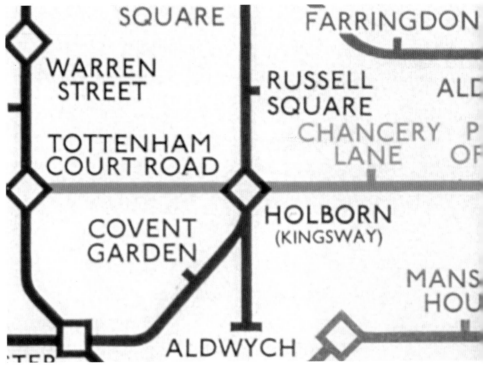

3

◄ **COCKFOSTERS AND THE PICCADILLY LINE** In the 1930s, the government invested a massive £4 million (US$6.6 million) in extending the overcrowded Piccadilly Line northwards from Finsbury Park to its current terminus at Cockfosters. The extension was completed in stages and ran throughout the period Beck was working on his new map. In July 1933, the final section was completed, just as the first fold-out maps based on Beck's design were released. The forward-thinking Beck included the new extension as though it had always been there.

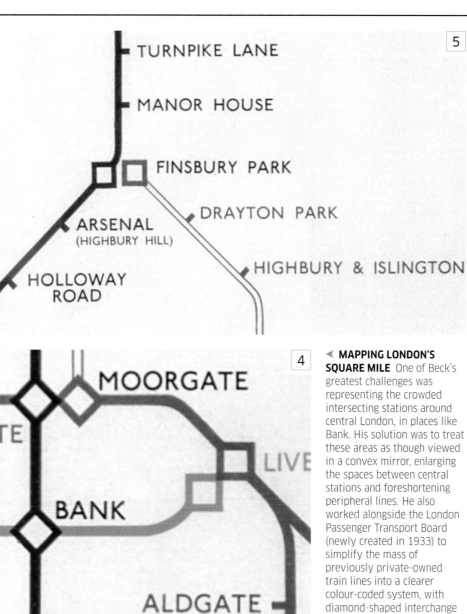

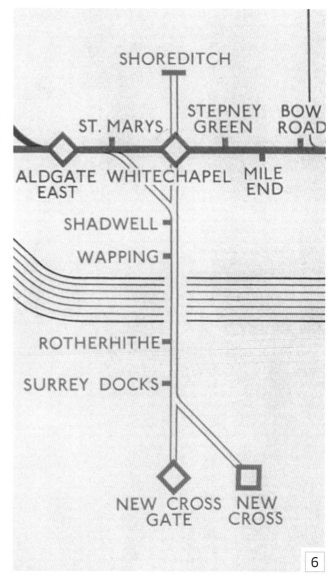

5 ◄ **FINSBURY PARK** Beck's map dealt with various line changes and connections to stations, most notoriously Finsbury Park in north London. Originally part of the private Great Northern and City line, by the early 1930s the proposed extension to the Piccadilly Line also involved Finsbury Park. Beck solved the problem by using two squares to identify the station's dual connection.

4 ◄ **MAPPING LONDON'S SQUARE MILE** One of Beck's greatest challenges was representing the crowded intersecting stations around central London, in places like Bank. His solution was to treat these areas as though viewed in a convex mirror, enlarging the spaces between central stations and foreshortening peripheral lines. He also worked alongside the London Passenger Transport Board (newly created in 1933) to simplify the mass of previously private-owned train lines into a clearer colour-coded system, with diamond-shaped interchange stations showing line changes. The diamonds have now gone, but Beck's colours remain.

6

ON **TECHNIQUE**

Beck was trained in drawing electrical circuits, a skill he used when redrawing the Underground, or "turning vermicelli into a diagram" as he put it. His early designs represent the train network like an electrical circuit board; what is essential for both is connecting one point to another, the distance travelled is not important. Later versions abandoned curves, adopting only straight or orthogonal (45-degree) lines, which were simple to use, but bore little relation to the world above.

➤ **One of Harry Beck's earlier sketches** shows the influence of his work with electrical circuits on the iconic London Underground map.

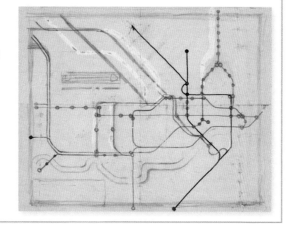

▲ **THE EAST LONDON LINE AND THE THAMES** Beck included just one physical feature above ground – the River Thames, represented by thin blue lines. This enabled passengers to orient themselves quickly on London's defining north-south axis. To the east, the river bisects the network's shortest line, the overland East London Railway (now part of the London Overground network), which runs between New Cross and Shoreditch, then the limit of the city but today two of London's fastest-growing districts.

Dymaxion Map

1943 ▪ PAPER ▪ 24CM × 39CM (9½IN × 1FT 3¼IN) ▪ BUCKMINSTER FULLER INSTITUTE, NEW YORK, USA

SCALE

BUCKMINSTER FULLER

With the outbreak of World War II in 1939, it became apparent that mapmaking could be used in the service of political, religious, and racial divisions. In 1943, the visionary American inventor and designer Buckminster Fuller decided to acknowledge the problems of projecting the spherical earth onto a flat surface by designing a map that offered a connected global world, and that stressed unity rather than difference.

His *Dymaxion Map*, named after his distinctive design ethic (see box below), used an icosahedron to create a terrestrial globe, which could be unfolded into a flat world map that looked like a piece of origami. Despite its unusual shape, it was more accurate in proportion than previous rectangular maps, which showed serious distortion, especially at the poles. Fuller's method proved that no map projection could accurately depict the whole globe. It also showed interconnected landmasses without political borders, reflecting his progressive belief in the need for global cooperation and sustainability. Fuller rejected cartographic orientations of "up" or "down", and instead created a radically democratic map that was more interested in how temperatures affected human development.

D   Y   M   A

N.   C.   STAT

R.   B

PUBL

COPY

TEMPERATURE

LAND			
WATER			

—58° F   —50° C   —49° F   —40° F   —40° C   —31° F   —22° F

BUCKMINSTER **FULLER**

1895-1983

Richard Buckminster Fuller was one of 20th-century America's great intellectual mavericks – an inventor, writer, architect, and designer.

Expelled from Harvard University, Fuller served in the US Navy during World War I. He worked on techniques for producing affordable, lightweight housing, the first of several innovative projects that came under his trademark term "dymaxion", a compound of three of his favourite concepts: dynamic, maximum, and tension. It described a series of increasingly ambitious projects that Fuller invented from the late 1920s, including a three-wheeled car, houses, and geodesic domes – stable, lightweight, spherical structures that influenced a generation of urban planners. Fuller's unorthodox ideas were based on his prescient belief in global sustainability and an environmental awareness of the fragility of what he called "Spaceship Earth".

A deck plan of the six and one half sextillion tons Spaceship Earth

BUCKMINSTER FULLER

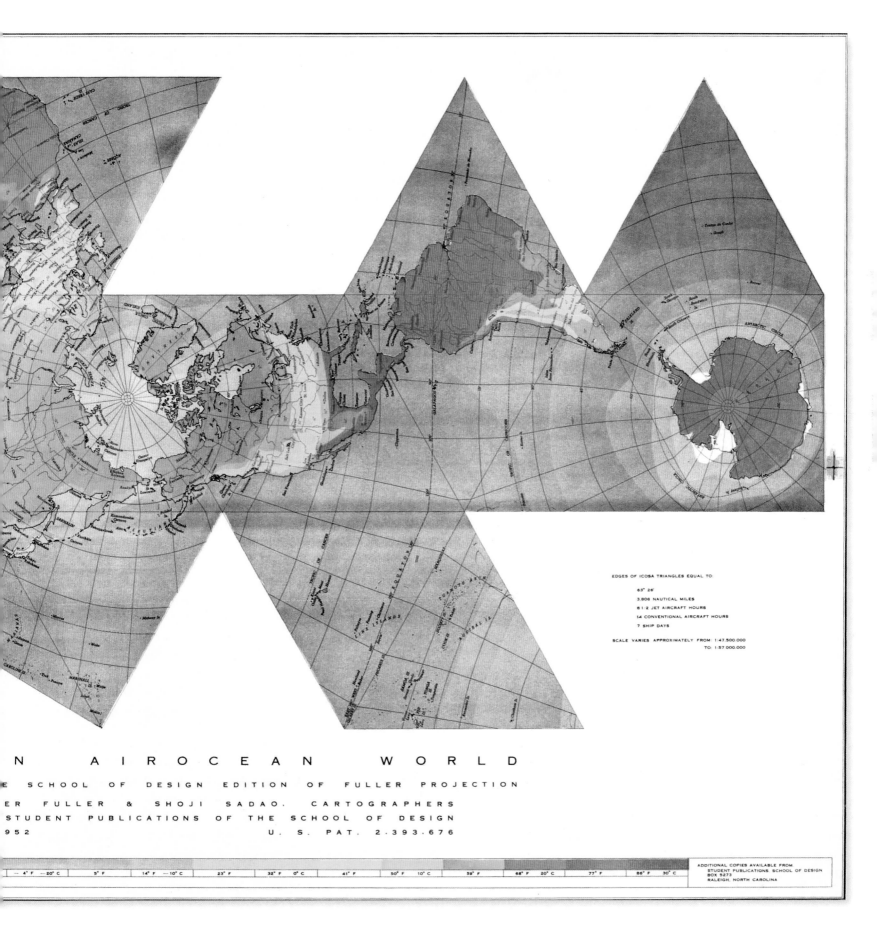

EDGES OF ICOSA TRIANGLES EQUAL TO:

63° 26'

3,806 NAUTICAL MILES

8 1/2 JET AIRCRAFT HOURS

14 CONVENTIONAL AIRCRAFT HOURS

7 SHIP DAYS

SCALE VARIES APPROXIMATELY FROM: 1:47,500,000
TO: 1:57,000,000

N A I R O C E A N W O R L D

E SCHOOL OF DESIGN EDITION OF FULLER PROJECTION

ER FULLER & SHOJI SADAO, CARTOGRAPHERS

STUDENT PUBLICATIONS OF THE SCHOOL OF DESIGN

952   U. S. PAT. 2,393,676

ADDITIONAL COPIES AVAILABLE FROM
STUDENT PUBLICATIONS, SCHOOL OF DESIGN
BOX 5273
RALEIGH, NORTH CAROLINA

— 4° F —20° C   5° F   14° F —10° C   23° F   32° F 0° C   41° F   50° F 10° C   59° F   68° F 20° C   77° F   86° F 30° C

Visual tour

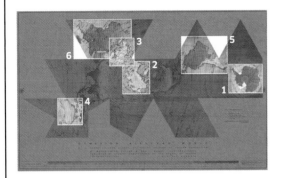

KEY

▶ **ANTARCTICA COMES IN FROM THE COLD**
Most world maps marginalize Antarctica, or in
Mercator's case, distort it (see pp.110–13). Even
polar projections invariably placed the North Pole
as their centre. Although Fuller's method situated
Antarctica in relative isolation, it offers a rare view of
the continent's size and shape. Current debates around
global warming make this image seem eerily prophetic.

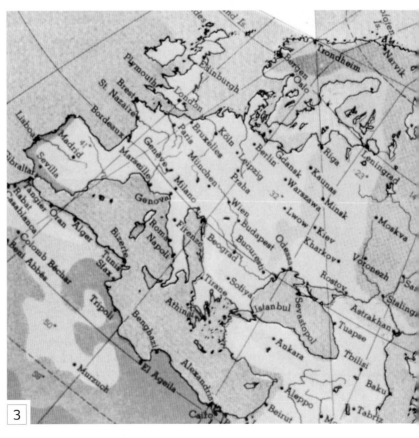

▲ **FROM ONE POLE TO ANOTHER**
One of Fuller's great beliefs was that
"up", "down", "north", and "south" were
all cultural constructs, and as a result
his map does not have a "right way
up". Although the North Pole sits
approximately in the middle, it has
no wider significance other than
to show a spiral-shaped region
of territorial interconnectivity.

▶ **MARGINAL EUROPE** Fuller decentres
Europe. Its political geography is no
longer seen as pivotal, with many of its
place names even written upside down.
Instead, Fuller reintroduces an age-old
interest in temperature zones, with
Europe lying within a temperate region.
This revision of Greek *klimata* (see p.43)
would have been recognizable to
Aristotle, Ptolemy, and even al-Idrīsī.

▼ DOWN UNDER? Having rejected orientation in terms of "up" and "down", Fuller's map changes assumptions made by the language of geography, including descriptions of Australia as "down under". Even the term "Antipodes" stems from Plato's explanation of "above" and "below", describing one place diametrically opposite, or "below", another. On Fuller's map, Australia is just another continent, cut free from age-old assumptions about its place in the world.

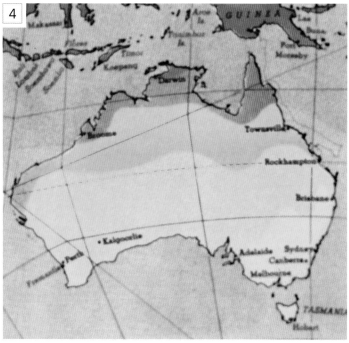

IN **CONTEXT**

Throughout history, most world maps projected the spherical globe onto straightforward shapes such as rectangles or ovals, which resulted in some form of distortion. Fuller took the radically different approach of using an icosahedron with twenty triangular faces, because it was the closest shape to a sphere, and therefore limited distortion when the earth's surface was projected upon it. The shapes and sizes of landmasses were preserved, but at the expense of producing an "interrupted" shape when the earthly icosahedron was flattened out into a discontinuous map.

▲ Fuller's map is shown here placed on to a three-dimensional icosahedron.

◄ SOUTH AMERICA
Having been the subject of so much imperial and colonial mapmaking since the late 15th century, South America is shown here as part of a ribbon connecting North America to Asia and then Africa in one continuous belt running left to right, rather than its usual orientation running north to south. Its distance from other continents, especially Africa, appears considerably distorted, but Fuller retains its shape and proportion.

► REDDEST AFRICA Compared with traditional maps, Africa looks "upside down", but Fuller's startling projection makes us realise that this is quite arbitrary. Its red hue is based on high temperature zones, although Fuller's interest was more in how social patterns of migration and economic activity were defined by the coldness of a region, as opposed to its heat.

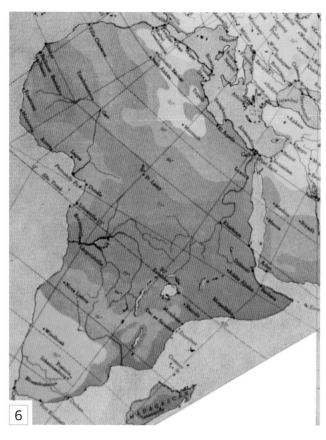

Lunar Landings Map

1969 ■ PRINT ■ 87CM × 89CM (34¼IN × 35IN) ■ GODDARD SPACE FLIGHT CENTER, GREENBELT, MARYLAND, USA

NASA

SCALE

LUNAR DATA

Distance from earth		
mean (miles)		238,900
(kilometers)		384,400
Diameters		
miles		2,160
kilometers		3,476
Temperatures		
sun at zenith		101°C to 130°C
night		−153°C
Velocity of escape		
miles per second		1.5
kilometers per second		2.4

GRAPHIC DATA

Position was established primarily from the measures of J. Franz and S.A. Saunder as compiled by D.W.G. Arthur and E. Whitaker in the Orthographic Atlas of the Moon, edited by Dr. Gerard P. Kuiper, 1960.

THE PHOTOGRAPHS IN THE MOSAICKED IMAGE OF THE MOON WERE SELECTED FROM PHOTOGRAPHY TAKEN AT Mc DONALD, MT. WILSON AND DU MIDI OBSERVATORIES.

With the advent of unmanned space travel, far more accurate maps of the Moon began to emerge than ever before. These are the culmination of centuries of selenography, the science of the study and mapping of the Moon, which stretches back at least to the early 17th century, when scientists such as Galileo started to map the lunar surface with the aid of telescopes. In 1651, the Jesuit priest and astronomer Giovanni Battista Riccioli not only mapped the Moon, but he also named many its features in Latin. The English translations of many of Riccioli's names, such as the Sea of Tranquility (*Mare Tranquillitatis*), are still used.

Throughout the 1960s, as it prepared the Apollo lunar landing programme, the American space agency NASA developed the Lunar Earthside Mosaic, a composite lunar lithographic map created from photographs taken from various observatories. First published in 1960 on a scale of 1:5,000,000, it used a projection (called orthographic) and was continuously updated, culminating in this 1969 edition on a scale of 1:2,500,000. Used to trace various subsequent Moon landings, it shows the enduring fascination of extra-terrestrial mapping.

ON **TECHNIQUE**

The lunar map photographs of the 1960s have been superseded by more sophisticated techniques. In 1994, the US *Clementine* lunar mission created a topographic map of the near and far sides of the Moon using lasers to capture height and surface relief, revealing previously hidden craters and basins. Since 2009, NASA's Lunar Reconnaissance Orbiter has been circling just 50km (30 miles) above the Moon, capturing data to make a three-dimensional map of its surface in order to locate possible future landing sites, and find potential lunar resources. The latest maps released have a resolution of 100m (330ft) per pixel.

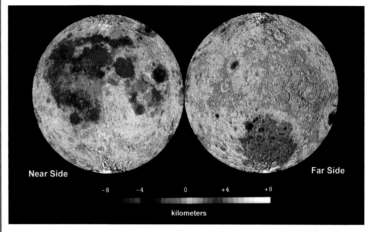

Near Side   Far Side

-8   -4   0   +4   +8
kilometers

▲ **The lunar probe** *Clementine* captured these images of the Moon, showing variation in topography using the coloured scale bar.

Visual tour

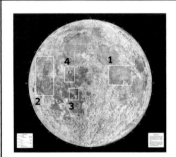

KEY

1

▲ **SEA OF TRANQUILITY** Riccioli mistook the Moon's vast, dark plains for seas, naming this one "Tranquility". It achieved immortality as the landing site of Apollo 11 in July 1969, with Neil Armstrong's words, "Houston, Tranquility Base here. The Eagle has landed".

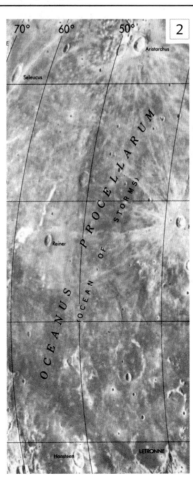

2

◀ **OCEAN OF STORMS** Riccioli referred to the larger lunar basins as oceans. At 4 million sq km (1½ million sq miles), this one is the Moon's largest. Astronomers speculate that it was created by the impact of a giant meteor. In November 1969, it was the landing site for Apollo 12.

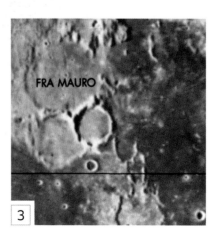

3

▲ **FRA MAURO HIGHLANDS** This crater and surrounding highlands, the landing site for Apollo 14 in 1971, was named after the 15th-century Venetian mapmaker (see pp.72–75). Fra Mauro's map showed both the Earth and the heavens.

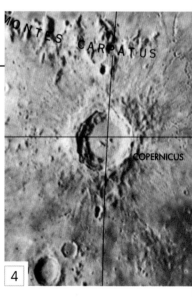

4

▲ **COPERNICUS CRATER** East of the Ocean of Storms lies a crater Riccioli named after Renaissance astronomer Nicolaus Copernicus. Perhaps Riccioli was wittily condemning Copernicus's heliocentric model of the universe to perpetual storms, or perhaps, as some historians have speculated, putting Copernicus on the lunar map was Riccioli's way of showing his tacit acceptance of his views.

Equal Area World Map

1973 ▪ PRINT ▪ 53CM × 82.5CM (1FT 8¾IN × 2FT 8½in) ▪ ODT MAPS, MASSACHUSETTS, USA

SCALE

ARNO PETERS

In 1973, Arno Peters, a German historian turned mapmaker, unveiled a remarkable new world map at a press conference in Bonn, then in West Germany. Peters claimed that his map offered equality to the world's colonized countries and to people living in what he called the southern "developing world". This included most of Africa, South America, and southeast Asia, covering 62 million sq km (24 million sq miles), in contrast to the countries of the dominant "developed" northern hemisphere, which cover just 30 million sq km (11½ million sq miles). Peters blamed Gerard Mercator's 400-year-old projection (see pp.110–13) for dominating world mapmaking from a Eurocentric perspective, which, he argued, "presents a fully false picture particularly regarding the non-white-peopled lands".

An advocate of social equality, Peters used an "equal-area" method, which prioritized fidelity to surface area. The result, he argued, was a map that was more accurate than Mercator's and that restored the size and position of developing countries. However, professional cartographers argued that the distortion of distances in Peters' map was more of a problem than the distortions in areas of other projections, and pointed out that no flat, rectangular world map could reproduce a spherical Earth without distortion. Initially, many religious, aid, and political organizations adopted Peters' map, distributing more than 80 million copies worldwide, and it has gone through various updated editions, including this one from 2014. However, it is rarely used today.

ARNO **PETERS**

1916–2002

Committed to issues of equality and social justice throughout his life, German historian and mapmaker Arno Peters caused controversy but raised important issues through his work.

Born in Berlin into a family of left-wing activists, Peters trained in film production techniques in the 1930s, before turning to history and writing a dissertation on film as a propaganda. Witnessing the horrors of World War II and his country's division afterwards profoundly affected his political views, and led him to work in East Germany as an independent scholar. In 1952, Peters published an innovative but controversial "synchronoptic" world history, which gave equal weight to non-Western history, and which he called "a map of time". This work inspired him to apply his interests to geography, and begin work on his world map. Following its publication in 1973, Peters also published a manifesto, *The New Cartography* (1983), and the bestselling *Peters Atlas of the World* (1989). Peters regularly revised his famous map and today, more than a decade since his death, it is still kept up to date.

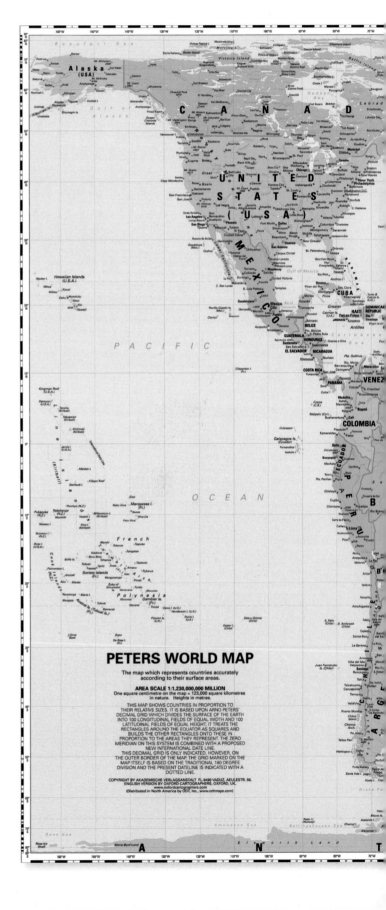

PETERS WORLD MAP

The map which represents countries accurately according to their surface areas.

AREA SCALE 1:1,230,000,000 MILLION
One square centimetre on the map = 123,000 square kilometres in nature. Heights in metres

THIS MAP SHOWS COUNTRIES IN PROPORTION TO THEIR RELATIVE SIZES. IT IS BASED UPON ARNO PETERS' DECIMAL GRID WHICH DIVIDES THE SURFACE OF THE EARTH INTO 100 LONGITUDINAL FIELDS OF EQUAL WIDTH AND 100 LATITUDINAL FIELDS OF EQUAL HEIGHT. IT TREATS THE RECTANGLES AROUND THE EQUATOR AS SQUARES AND BUILDS THE OTHER RECTANGLES ONTO THESE IN PROPORTION TO THE AREAS THEY REPRESENT. THE ZERO MERIDIAN ON THIS SYSTEM IS COMBINED WITH A PROPOSED NEW INTERNATIONAL DATE LINE.
THIS DECIMAL GRID IS ONLY INDICATED. HOWEVER, ON THE OUTER BORDER OF THE MAP THE GRID MARKED ON THE MAP ITSELF IS BASED ON THE TRADITIONAL 180 DEGREE DIVISION AND THE PRESENT DATELINE IS INDICATED WITH A DOTTED LINE.

COPYRIGHT BY AKADEMISCHE VERLAGSANSTALT FL-9490 VADUZ, AEULESTR. 56. ENGLISH VERSION BY OXFORD CARTOGRAPHERS, OXFORD, UK. www.oxfordcartographers.com (Distributed in North America by ODT, Inc. www.odtmaps.com)

The landmasses are somewhat reminiscent of
wet, ragged, long winter underwear hung out
to dry on the Arctic Circle

ARTHUR ROBINSON, AMERICAN CARTOGRAPHER

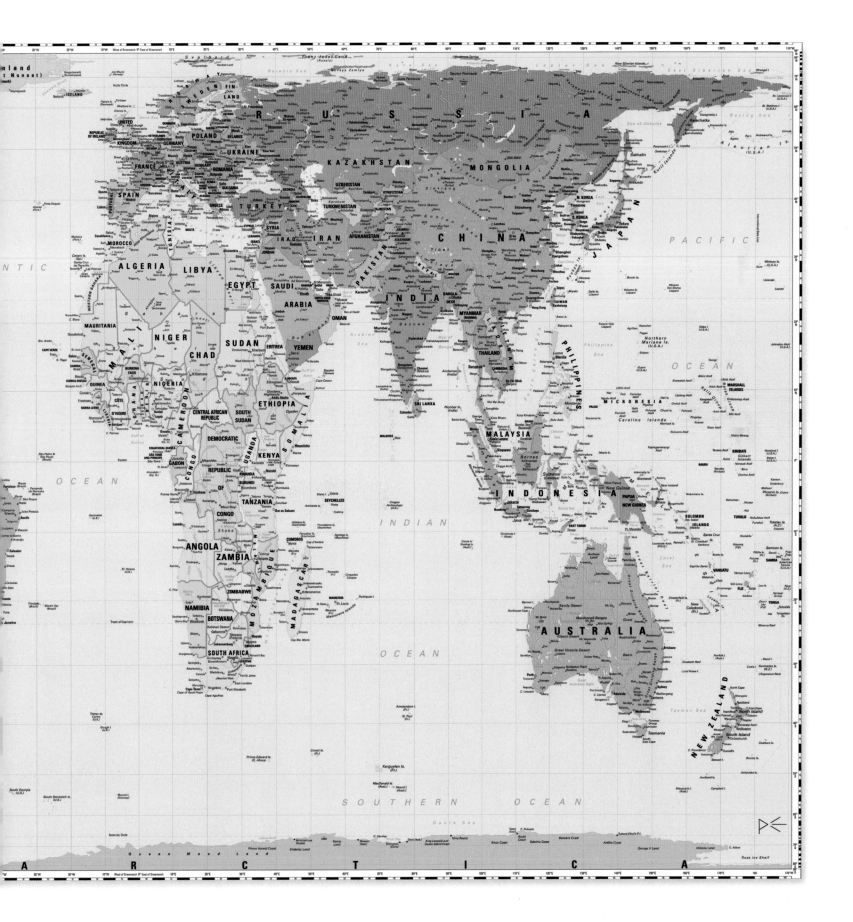

Visual tour

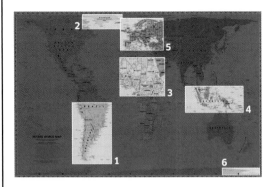

KEY

▼ **GREENLAND** Peters pointed out that Mercator's projection distorted the sizes of landmasses. Greenland covered 2.1 million sq km (811,000sq miles), but on Mercator maps it dwarfed China's 9.5 million sq km (3¾ million sq miles). Peters' projection corrected this, although the distortion on the map's northern and southern extremities gave the country a strangely flat shape.

▶ **SOUTH AMERICA** Covering 17.8 million sq km (nearly 7 million sq miles), South America had traditionally always seemed much smaller than other continents, including Europe. One of the most striking aspects of the Peters map was the way it elongated the American and African continents, much to the consternation and amusement of many professional cartographers.

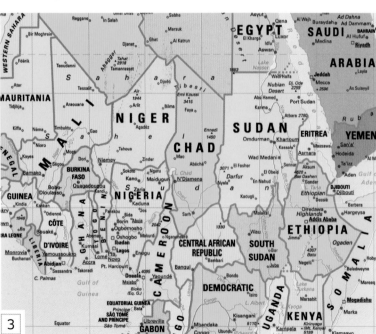

▲ **CENTRAL AFRICA** Despite covering 20 per cent of the Earth's land surface, Africa has always been diminished on modern Western maps, and often shown at a smaller scale than Europe and North America. Peters was eager to compensate for this, but he went too far. Critics noted that his map showed Chad and its neighbour Nigeria twice as long as they should be, even according to Peters' own equal-area formulation.

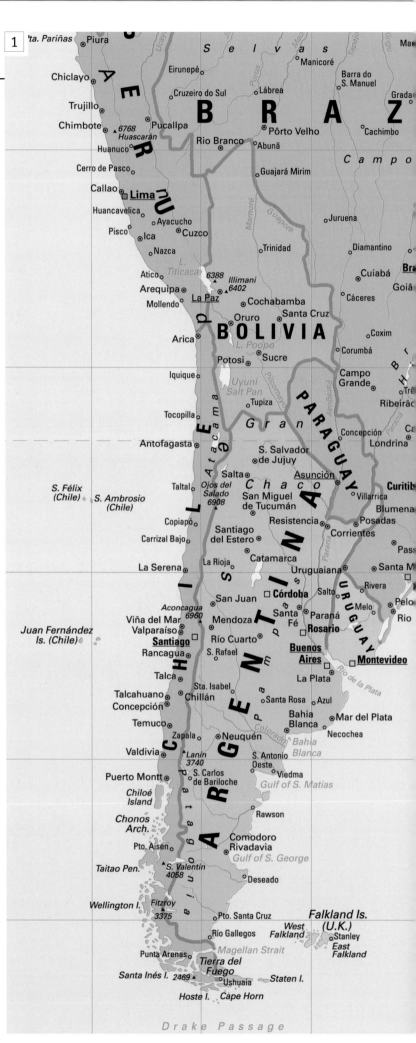

4

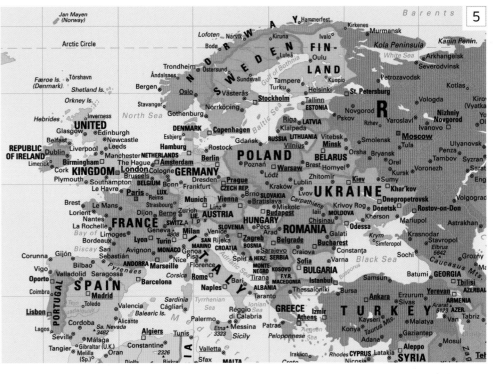

5

◀ **INDONESIA** As the centre of the spice trade, Indonesia has always been of great interest to European mapmakers. Peters saw it as part of the developing world, either marginalized or distorted by Western geography. Unfortunately, his mapping of Indonesia introduced new errors, depicting it at twice its north-south height and half its east-west breadth.

◀ **EUROPE** Peters' objection to most Western projections was their exaggeration of the size and centrality of Europe. Covering just 9.7 million sq km (3¾ million sq miles) and with the UK taking up only 0.16 per cent of the Earth's land surface, Europe is smaller than any other continent, although Western maps and atlases prior to Peters had not reflected this.

▼ **ANTARCTIC PENINSULA** Despite attacking Mercator's projection, Peters chose an orientation and rectangular shape that reproduced many of the Flemish mapmaker's errors and obvious distortions. He depicts the Antarctic Peninsula just like Mercator, stretching it to infinity east to west, because Peters' projection imagines the Earth as an unrolled cylinder.

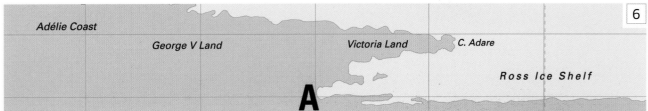

6

ON **TECHNIQUE**

Peters used a technique called an orthographic equal-area projection to construct his map. This involved treating the globe like a cylinder, which was then unrolled to create a rectangle, which, Peters insisted, was the best way to represent the Earth on a flat surface. He took 45 degrees north and south as his two standard parallels – places at which there is minimal distortion – and then calculated the surface area of the landmasses on his map, creating an "equal-area" depiction of countries relative to each other. Peters claimed this was a revolutionary method, unprecedented in its accuracy, although it had actually been invented in 1855 by the Reverend James Gall (1808–95), a Scottish evangelical minister and amateur mapmaker.

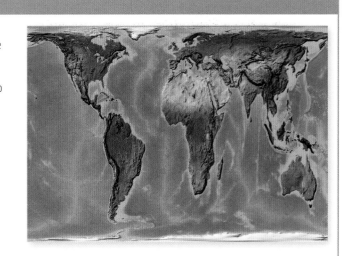

▶ **This equal-areas map** is based on both James Gall's and Arno Peters' calculations and is known as the Gall-Peters projection.

World Ocean Floor

1977 ▪ PAPER ▪ 56.7CM × 97.4CM (1FT 10¼IN × 3FT 2¼IN)
▪ MARIE THARP MAPS, NEW YORK, USA

SCALE

MARIE THARP AND BRUCE HEEZEN

Women's involvement in map history is notoriously limited. However, there are several honourable exceptions, one of whom is geologist Marie Tharp, who began studying the ocean floor at Columbia University in the US in 1948. Over the next 20 years, she compiled data pointing to the theory that continental drift – the movement of the Earth's continents across the ocean bed – was occurring due to the shift of tectonic plates (large pieces of the Earth's crust). Unfortunately, as a woman, Tharp was not allowed on board the exploratory sea voyages that measured the oceans' depth and contours; so, she worked with collaborator Bruce Heezen to assess the resulting data. The result was the first map of the ocean floor, showing a dramatic, mountainous underwater landscape that convinced the scientific community of the reality of continental drift.

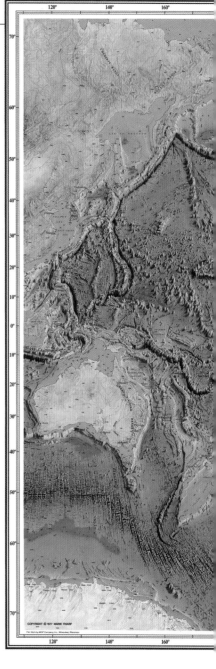

MARIE **THARP**

1920–2006

Michigan-born Marie Tharp began work as a geologist and cartographer at the Lamont-Doherty Earth Observatory at Columbia University in 1948, where she met her lifelong collaborator, Bruce Heezen.

Tharp and Heezen's first "physiographic" maps appeared in 1957, and they published the complete map of the world's ocean floor in 1977; that same year, Heezen died while on a research voyage near Iceland. Tharp continued her work at Columbia until she retired in 1983.

Visual tour

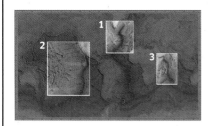

KEY

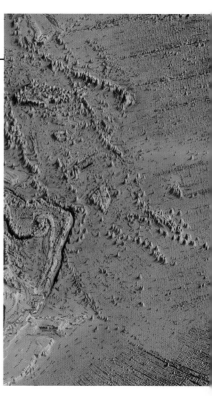

▶ **THE MID-ATLANTIC RIDGE** What appears to be a mountain range known as the "Mid-Atlantic Ridge" runs down the Atlantic from Iceland to the Canaries. However, Tharp also noticed a rift valley within the range, which could only be a seam in the Earth's crust, a point where tectonic plates collided causing "drift". It was a sensational discovery that shook the geological world, which was primarily composed of "fixists" who dismissed "drifters" as eccentric.

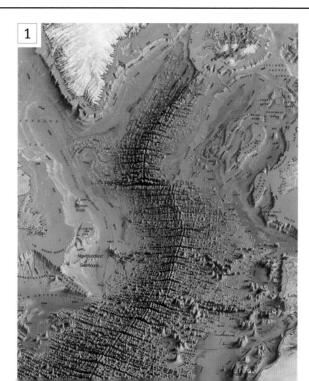

1

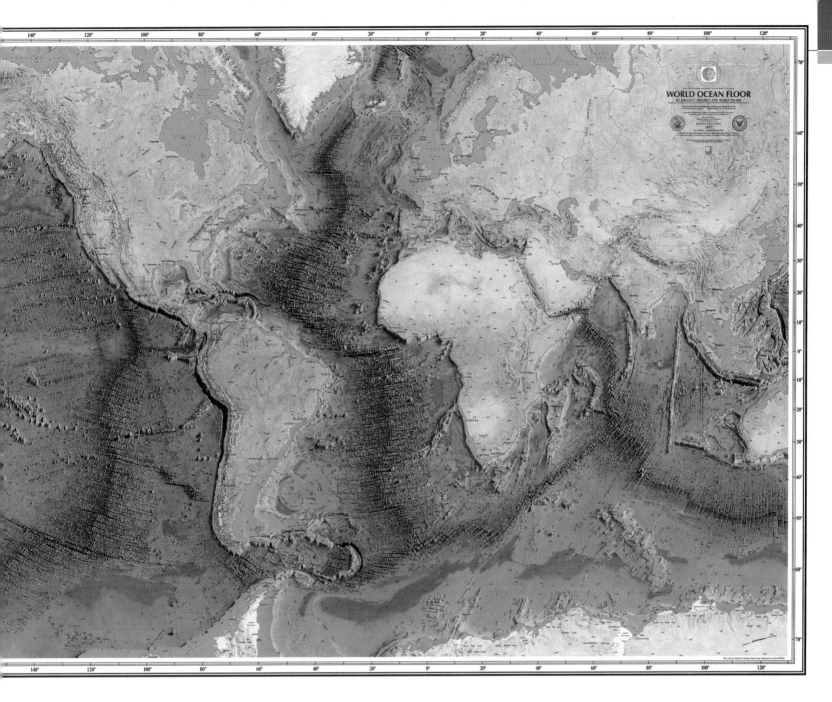

WORLD OCEAN FLOOR
BY BRUCE C. HEEZEN AND MARIE THARP

1977

◄ **PACIFIC DRIFT** As Tharp expanded her data around the world, she found that rift valleys such as the one in the Atlantic were a feature of all ocean floors. The so-called "East Pacific Rise", off the west coast of America, appears on Tharp's map like an enormous crack where it meets several tectonic plates, including the Earth's largest, the Pacific plate.

▶ **THE EAST AFRICAN RIFT SYSTEM** This was another dramatic rift shown by Tharp, running for over 6,400km (4,000 miles) and stretching 64km (40 miles) wide from the Arabian Peninsula through the Indian Ocean. Working with the artist Heinrich Berann, Tharp created a three-dimensional map as visually powerful as any in cartographic history.

3

ON **TECHNIQUE**

Tharp assessed sonar readings taken by the exploratory ship *Vema*, which measured depth by bouncing sound waves off the ocean floor. She plotted depth measurements on massive paper sheets, building up a three-dimensional profile of the ocean's surface. This allowed her to identify the seams in the Earth's crust, which suggested continental drift. Tharp then began compiling her "physiographic" ocean floor map, with its distinctive surface relief (shown here).

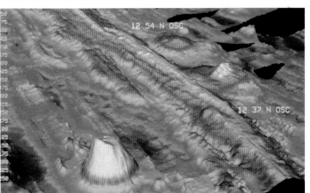

▲ **This image shows the East Pacific Rise** on the sea bed of the Pacific Ocean, obtained using sonar readings.

Mappa

1989 ▪ EMBROIDERY ON LINEN ▪ 2.3M × 4.6M (7FT 6½IN × 15FT 1IN)
▪ MUSEUM OF MODERN ART, NEW YORK, USA

ALIGHIERO BOETTI

SCALE

Artists have always been fascinated by the visual power of maps, and none more famously than Alighiero Boetti in his celebrated series, "Mappa". In the late 1960s, Boetti became interested in maps, drawing the outline of Palestine and the "occupied territories" as a man-made shape which, once copied, could be labelled "art". Later, in 1971, Boetti made the first of many visits to Afghanistan. He was fascinated by the embroidery methods and

The world is shaped as it is, I did not draw it; the flags are what they are, I did not design them. In short I created absolutely nothing

ALIGHIERO BOETTI

materials of local weavers. Working through intermediaries, he commissioned these craftswomen to embroider a series of world maps, providing no instructions as to design and colour beyond the map itself. Boetti was

interested in making art that was created by different producers, without collaboration or discussion, leaving many elements to chance and accident – such as the colour of the oceans on the maps, which was interpreted in different ways by the weavers. He also wanted to make art that was removed from the process of invention; the act of copying a map was a perfect example of this.

A world in context

Over time, the series took on different meanings. As a conceptual artist who understood map projections, Boetti knew that the idea of showing the world "as it really is" was futile. He was more interested in showing how the gulf between western cartography and Afghan craftsmanship could be bridged by producing an object that united them. Each *Mappa* also registered the flux of global politics: by using flags to demarcate national boundaries, Boetti was able to trace changes to the political world map over time, especially in the maps' war-torn country of origin, Afghanistan.

ALIGHIERO **BOETTI**

1940-1994

One of Italy's first and foremost conceptual artists, Alighiero Boetti was born in Turin, Italy, and studied engraving in Paris, France.

Boetti pioneered the *Arte Povera* ("Poor Art" – see p.235) movement from his home town in the late 1960s. Strongly influenced by the artist Marcel Duchamp, Boetti's early works were highly intellectual, minimal, yet playful pieces using plastic, industrial fabric, and lighting to examine the boundaries between art and life. He was also fascinated by letters, numbers, and games, anticipating his interest in maps, which developed during periods of travelling in Afghanistan in the 1970s. Boetti also founded the One Hotel in Kabul, which became his Afghan base and was where he conceived the idea for his famous Mappa series of artworks.

Visual tour

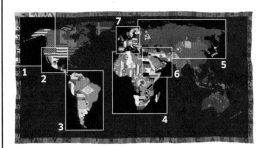

KEY

> **WRITTEN BORDER** The left border, in Farsi, reads, "The star-filled sky of Kabul is the same as the plain desert"; to the right, the text reads "The fresh rain drops from the Kabul sky onto the room of Alighiero e Boetti". At the top and bottom, in Italian, it reads, "During your life of wandering my brother, always keep your eye fixed firmly on the doughnut and never the hole".

▲ **MEXICO** Boetti ordered the latest maps and flags from a London cartographer, which were then despatched to the Afghan weavers, many of whom had never seen world maps or national flags before. This, as Boetti desired, introduced an element of chance into the creative act, as in the bold colour contrast of the Mexican flag against the US "Stars and Stripes".

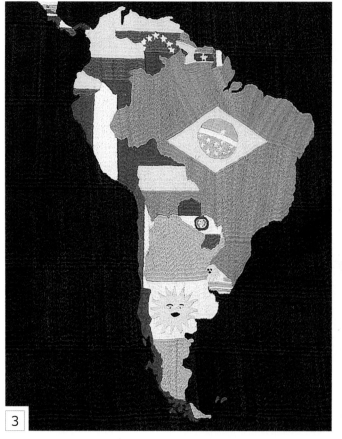

▲ **SOUTH AMERICA** Boetti experimented continually with different map projections, acknowledging contemporary debates about their political prejudices. Initially he used the Mercator projection, but the shape and size of South America, shown here, reveals that by the late 1980s he favoured the American cartographer Arthur Robinson's widely used projection of 1963.

▶ **AFRICA** Demonstrating the speed of political change in postcolonial Africa, Mozambique flies its new post-1975 independence flag, while Angola's flag evokes the Soviet hammer and sickle, following independence from Portuguese colonial rule the same year. Just below it, Namibia is completely blank due to the civil war of 1966–90 that left it flagless.

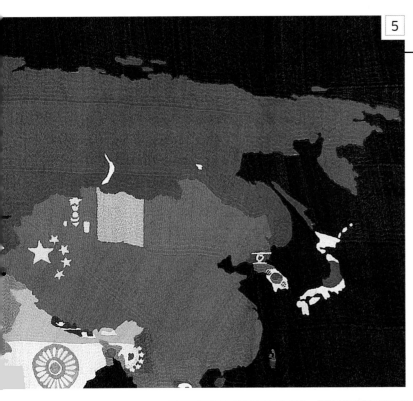

5

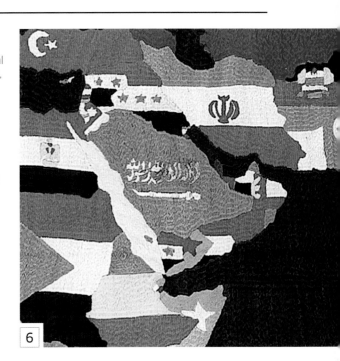

6

◄ RUSSIA AND AFGHANISTAN

The map captures a critical moment in the late 1980s, when the red and yellow hammer and sickle of the Soviet Union still dominated central Asia. Afghanistan was also being ruled by a doomed Communist republic – represented by its second flag (black, green, and red), used between 1987 and 1989. Both states disappeared within three years, replaced with alternative ideologies, and represented by different flags.

7

▲ MIDDLE EAST In the late 1980s,

the Middle East, including Iraq and Iran, was one of the more stable international regions. Looking at the map in the aftermath of the 21st century's Gulf Wars and Arab Spring emphasises Boetti's point about the ceaselessly fluid "fabric" of the political world.

◄ EUROPE Closer inspection of Europe

betrays tensions simmering just under its apparently harmonious surface. As Soviet Communism began to unravel with the fall of the Berlin Wall in 1989, down would come the flags of Yugoslavia (bottom right) and up would rise new ones in the Baltic States, still coloured here in an undifferentiated block of Soviet red.

IN **CONTEXT**

Boetti rose to prominence as part of the Italian *Arte Povera* movement of the 1960s, a post-war reaction to consumerism and the grand painting tradition of American Abstract Expressionism. Instead, the movement embraced simple, everyday objects, including stones, wood, fabric, and plastic, often minimally arranged to avoid suggestions of too much artistic endeavour. The body also became important as a place of experimentation, as artists like Boetti sought a more direct and playful language of artistic expression that went beyond traditional painting and sculpture.

➤ **Alighiero Boetti's sculpture** *Me Sunbathing in Turin 19 January 1969* on display in the Tate Gallery, London, UK.

Cartogram

2008 ▪ DIGITAL ▪ UNIVERSITY OF SHEFFIELD, SHEFFIELD, UK

WORLDMAPPER

As statistical data on global issues becomes more complex and digital technologies generate increasingly innovative and versatile mapping techniques, the cartogram has become one of the most important recent cartographical developments. A cartogram uses a single variable subject that can be measured statistically, such as population or immigration, which is then mapped on to land areas to convey an image of its proportional distribution.

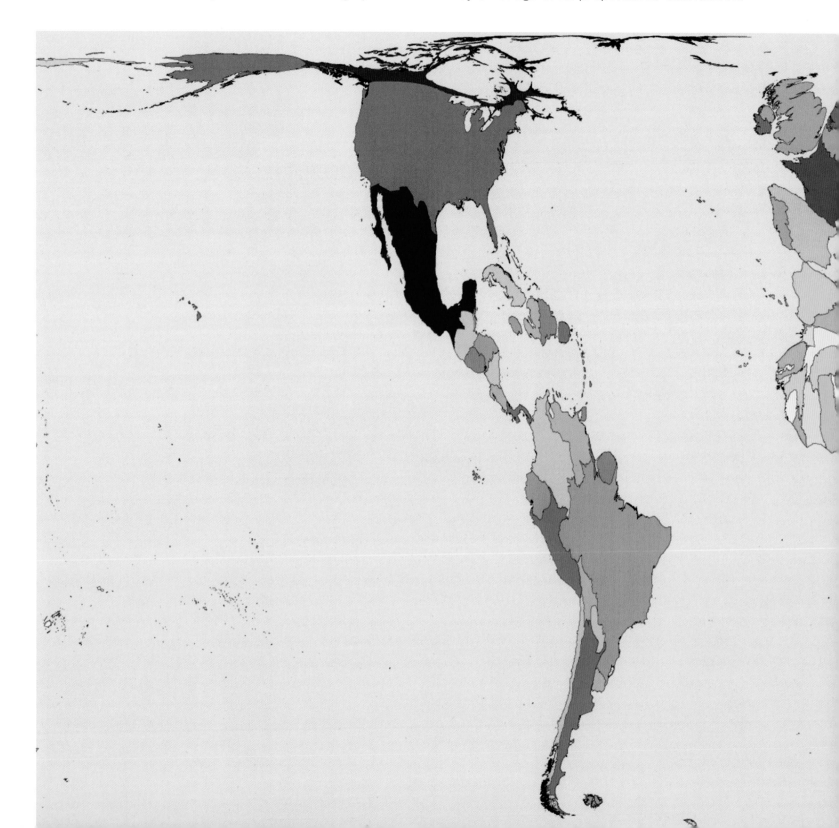

The British academic Danny Dorling is one of the method's most innovative practitioners. He worked as part of a team called Worldmapper, which used data gathered from a variety of organizations including the United Nations (UN), the World Health Organization (WHO), and the World Bank, to map subjects as diverse as world population (shown here), transport, poverty, health, war, and prostitution.

Rather than using traditional colour or shading, these maps inflate or shrink countries according to each subject. This cartogram shows the distribution of the Earth's estimated population of six billion in 2000.

DANNY **DORLING**

1968–

Currently the Halford Mackinder Professor of Geography at Oxford University, Danny Dorling was trained as a geographer and has held academic posts at several UK universities.

Dorling is a renowned social geographer who has published extensively. His work is characterized by the use of mapping techniques to visualize a variety of social and demographic statistical information, usually with a strong political and moral message, on subjects such as poverty, mortality, and housing.

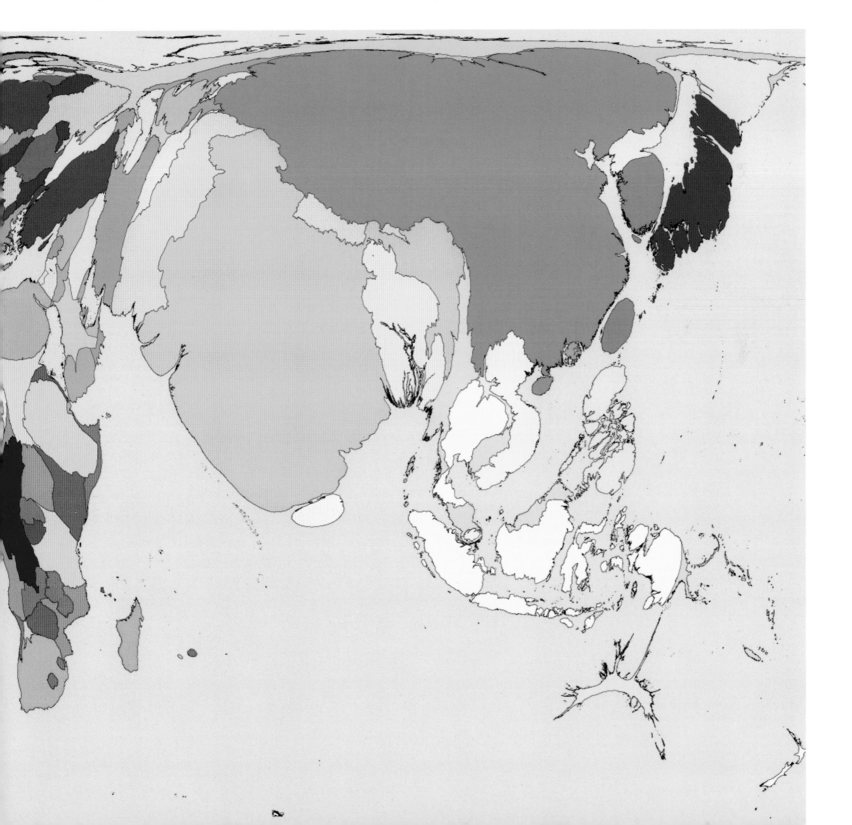

Visual tour

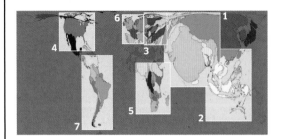

KEY

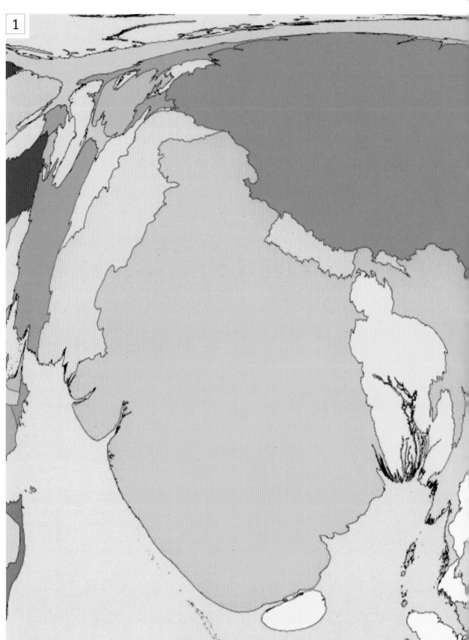

1

▶ **INDIA** Although marginalized on many historical maps, India is prominent on most geopolitical maps today, including population cartograms such as these. With over 1.2 billion inhabitants, it is the world's second most populous country, after China. However, India's population is predicted to eclipse China's by 2025 and by 2050, it is estimated that India will have 1.6 billion inhabitants.

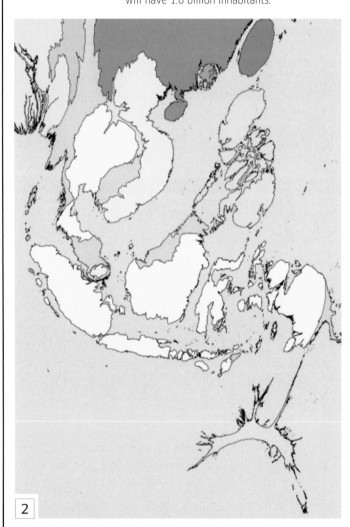

2

▲ **SOUTHEAST ASIA** Over 4.2 billion people live in the Asia-Pacific region, representing around 60 per cent of the global population. However this figure obscures complex demographic variations: population growth rates have actually dropped in the region to just 1 per cent, due to declining birth rates and lower death rates. The region's greatest challenge is an estimated threefold increase in the number of people aged over 65 – to an estimated 1.3 billion – by 2050.

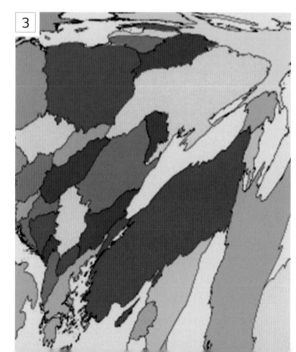

3

▶ **RUSSIA** One of the cartogram's most surprising results shows Russia – a geopolitical giant – suffering dramatic shrinkage, thereby revealing some of the country's problems. With just eight people per square kilometre (about 21 per square mile), it is one of the world's most sparsely populated countries, and in recent decades mortality rates have been extremely high while birth rates have been low. Although now stable, some estimates predict that Russia's population will contract to just 107 million by 2050.

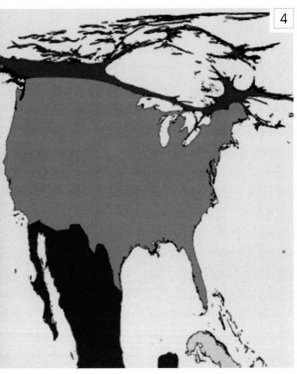

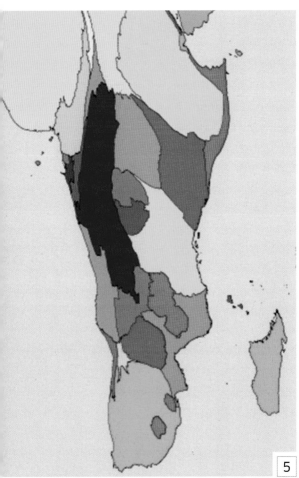

◄ NORTH AMERICA
Like Russia, the United States does not look as large or imposing as one might expect. With a modest population density of 34 people per square kilometre (or around 88 per square mile), its current population of 312 million remains relatively stable due to falls in immigration. Nevertheless, by 2050, the population is expected to grow to nearly 400 million, an increase of 28 per cent.

ON **TECHNIQUE**

Worldmapper's team created a series of high-impact cartograms by applying differential equations to a single global variable, such as population data. They used a mathematical model, driven by elementary physics, to "warp" the conventional world projection and adjust its proportions according to the chosen variable. The results are dramatic, conveying statistics in a radical, visual way that an ordinary table could not hope to achieve.

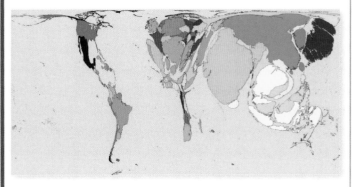

▲ **This "mopeds and motorcycles"** cartogram reveals in which countries the ownership of motorcycles is most prominent.

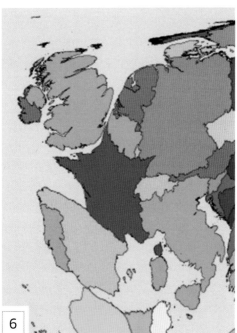

▲ **UNITED KINGDOM** With a population of 63.7 million, the United Kingdom is the third largest nation in Europe, behind Germany and France. With 256 people per square kilometre (around 666 per square mile), it also has one of the world's highest population densities, hence its prominence.

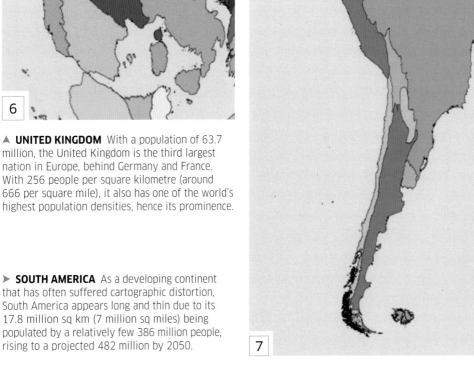

▲ **AFRICA** Relations between Africa's population and land area are striking: Sudan is Africa's largest country, although its population of 35 million is dwarfed by the much smaller Nigeria (177 million). By 2050, the population of sub-Saharan Africa, one of the world's poorest regions, is predicted to double in size to 2.4 billion people.

▶ **SOUTH AMERICA** As a developing continent that has often suffered cartographic distortion, South America appears long and thin due to its 17.8 million sq km (7 million sq miles) being populated by a relatively few 386 million people, rising to a projected 482 million by 2050.

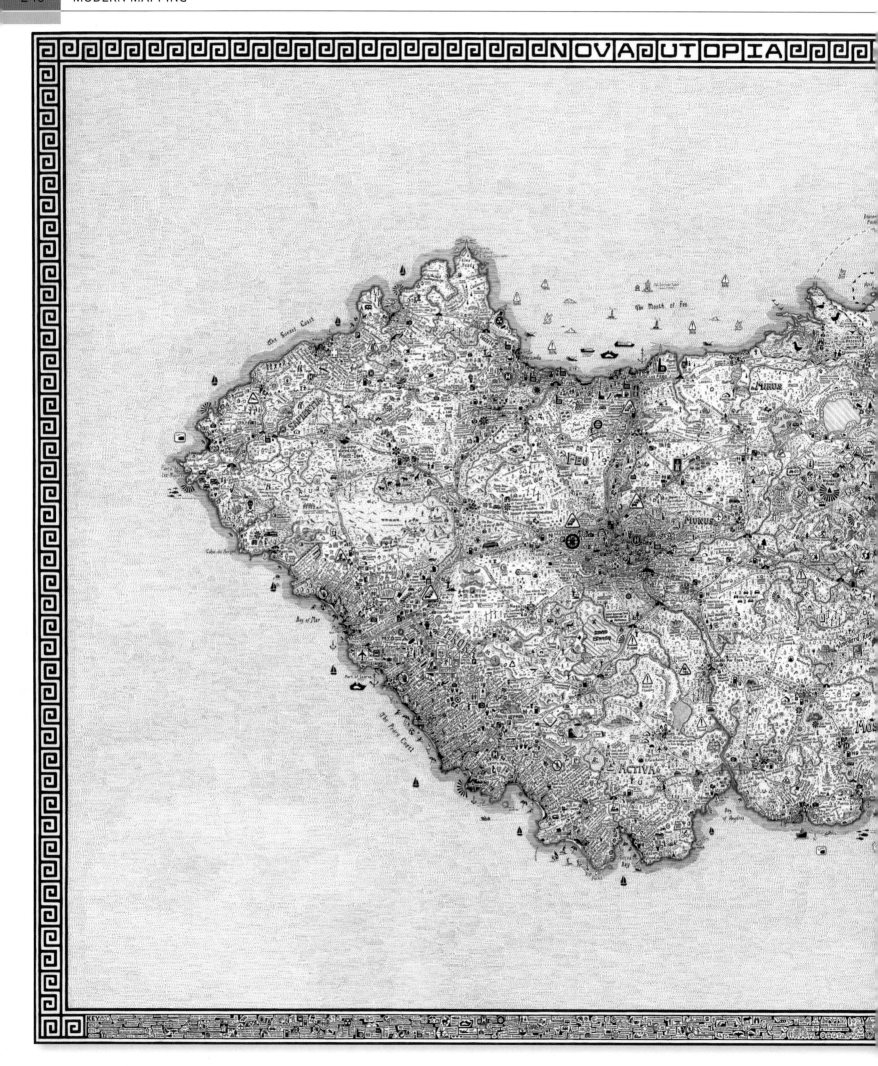

Nova Utopia

2013 ▪ DIGITAL PRINT ▪ 1.34M × 1.72M
(4FT 4½IN × 5FT 7½IN) ▪ TAG FINE ARTS, LONDON, UK

SCALE

STEPHEN WALTER

One of the wittiest and most penetrating recent attempts to map the concept of Utopia is found in Stephen Walter's *Nova Utopia*. Rather than drawing his own conception of an ideal state, Walter bases his work on Thomas More's original idea of Utopia (see pp.94–95), treating it as if it were a real place. His map describes an island transformed by a capitalist revolution in 1900, when a group of "Entrepreneurs" crushed the "Utops" on 23 April 1900. This may be an allusion to St George's Day, suggesting that, like More's *Utopia*, Walter's map is also a commentary on modern Britain. The triumph of private enterprise turned the island into a wealthy tourist destination, known as the "Leisure Island".

Walter asks if Utopia is still a valid idea, and shows how easily its ideals can be compromised. With compulsively drawn detail he depicts the darker, dystopian aspects of contemporary life usually omitted from maps. Yet this work, he says, also "glories in landscape, semiotics, etymology, and the intricate details of life".

> *Nova Utopia* sits somewhere between the wonderful, the beautiful, the entertaining, the rich, the sublime, and the ridiculous

STEPHEN WALTER

STEPHEN **WALTER**

1975–

A graduate of London's Royal College of Art, British artist Stephen Walter is renowned for prints and drawings exploring the concept of place, usually mapped in intricate, almost obsessive detail.

Much of Walter's work documents his home town, London. His print *Subterranea* (2012) shows a buried, underworld London including disused underground railway lines, sewers, burial sites, historical trivia, and ghost stories, while his sprawling work *The Island* is a humorous, semi-historical map of the city, full of stereotypes, local knowledge, personal anecdotes, and little-known facts. His art is characterized by its wry, informal tone, and it often refers curious readers to books and websites for more information. Walter has exhibited his work in cities including London, Berlin, Sydney, San Francisco, and Tel Aviv, and has also designed book and album covers.

Visual tour

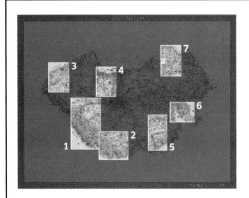

KEY

▶ **THE PRORA COAST, THE TOURIST'S DREAM**
Taking its name from a Nazi-designed beach resort, Prora is a satirical vision of the despoliation caused by modern mass tourism and overdevelopment. The area features commercialized seaside towns such as El Dorado, which is named after a mythical golden city, and also a much-derided British soap opera.

▲ **SPORT IN ACTIVA** An area "popular with the young and sporty", Activa also includes a red light district and areas inhabited by "locals that can no longer afford to live by the coast".

▶ **"AUTHENTIC" UTOPIA IN SAPIENTIA** From the Latin for "wisdom", Sapientia retains the vestiges of traditional utopian life, and is represented as one vast retirement home.

▼ **FEO, THE HOME OF NOVUS UTOPOS** Containing Novus Utopos – held by the locals to be the country's capital – the region of Feo (meaning "ugly" in Spanish) is a dilapidated industrial wasteland. Here, Walter observes with his acerbic political wit, "gentrification may well be on its way".

▼ **GETTING AWAY FROM IT ALL IN MOSRIS** The island's traditional rural retreat of Mosris has not escaped the impact of commercialization. Walter advises that "pre-booking your visits and transport is strongly advised".

IN **CONTEXT**

Walter's map of Utopia is part of a long artistic tradition, going back to Ambrosius Holbein's 1518 woodcut (see pp.94–95). *Nova Utopia*'s "aesthetic template" is a Renaissance map of Utopia made by the famous Flemish mapmaker Abraham Ortelius in 1598, which was directly inspired by Thomas More's *Utopia* (1516). Mapmakers and artists such as Walter are interested in Utopia as a graphic "no place", an invented world upon which they can project a range of contemporary hopes and fears.

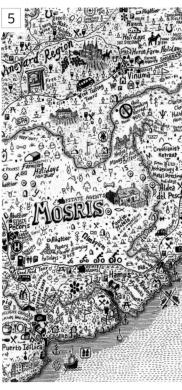

▲ **The title page** of Thomas More's *Utopia*, 1516.

▲ **FILTHY RICH IN FLOSRIS** Many people's utopian aspirations could be encapsulated in the wealthy retreat of Flosris, the playground of the super-rich, with its yachts and villas – it even includes "Paradise". However, it also has its dystopian dimensions: there are cosmetic surgery resorts, swingers' party houses, and trespassers are liable to prosecution.

▲ **AN ALTERNATIVE IN SACRUM** A refuge for alternative cultures opposed to privatization, Sacrum offers one ray of hope in Walter's new Utopia. It boasts ecological awareness and organic initiatives. Walter notes that this region has become more popular recently, in contrast to "the island's more mainstream tourist traps".

Google Earth

2014 ▪ DIGITAL ▪ CALIFORNIA, USA

GOOGLE

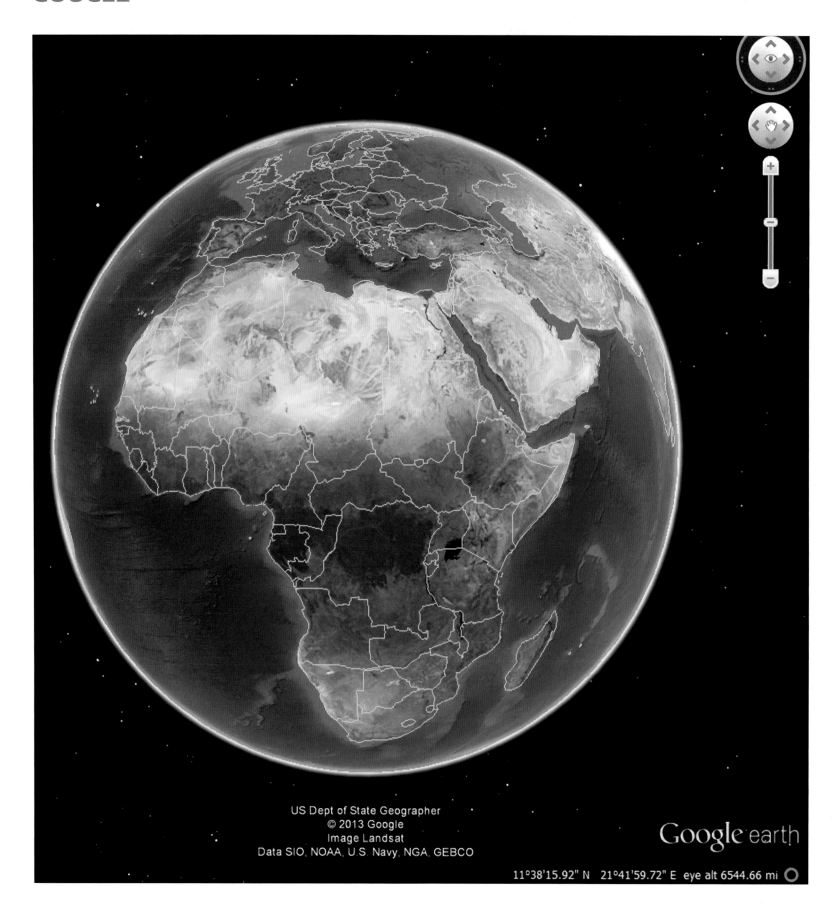

US Dept of State Geographer
© 2013 Google
Image Landsat
Data SIO, NOAA, U.S. Navy, NGA, GEBCO

Google earth

11°38'15.92" N   21°41'59.72" E   eye alt 6544.66 mi

The internet revolution of the 1990s transformed the way in which humans communicate, gather information, and even do business. The search engine Google currently dominates the market in "geospatial" applications – geographical data combined with computer software. In 2005, the company launched Google Earth, an application that covered most of the world and beyond, which enabled users to zoom from outer space to their own home, and back again, in seconds. Google Earth is pushing traditional definitions and even functions of maps: it allows users to turn political boundaries on or off, view terrain in 3D, and stream video content. Google estimates that a third of its internet searches require geographical information, and Google Earth, along with its Google Maps application, is believed to be used by over half of all mobile phone users. However, the scope of Google Earth and the information it offers has led to criticism for monopolization, infringing privacy laws, and even endangering global security.

▲ **Users can zoom in** on any part of this map to find details of local streets and businesses.

> The largest consumers of **map data** in the future will be **mobile**

BRIAN MCCLENDON, VICE PRESIDENT OF ENGINEERING, GOOGLE MAPS

Visual tour

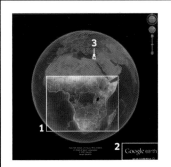

KEY

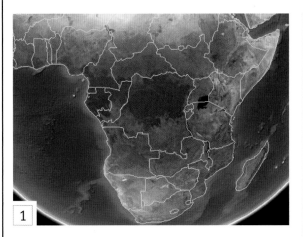

1

▲ **POLITICAL BOUNDARIES** Various tools allow users to personalize their maps: they can turn political boundaries on and off, as shown here in Africa, and even overlay older historical maps on to present day ones.

2

◀ **DISTANCE FROM EARTH** Google Earth's initial view shows the planet spinning in space, as though viewed from an astronaut's perspective. The altitude shown for the Earth is approximately that of a satellite orbiting the planet. As you zoom in, the altitude becomes lower.

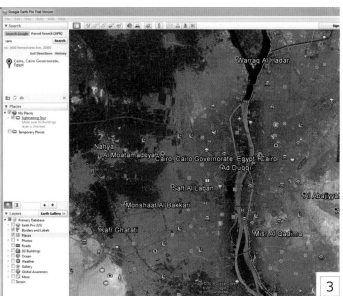

◀ **CAIRO** Users can switch easily from Google Earth to Google Maps, as shown here with Cairo. Zooming in, increasing levels of clickable features appear, from physical geography to buildings, business, shops, restaurants, and hotels. Google receives advertising revenue from businesses that pay to appear on the map.

3

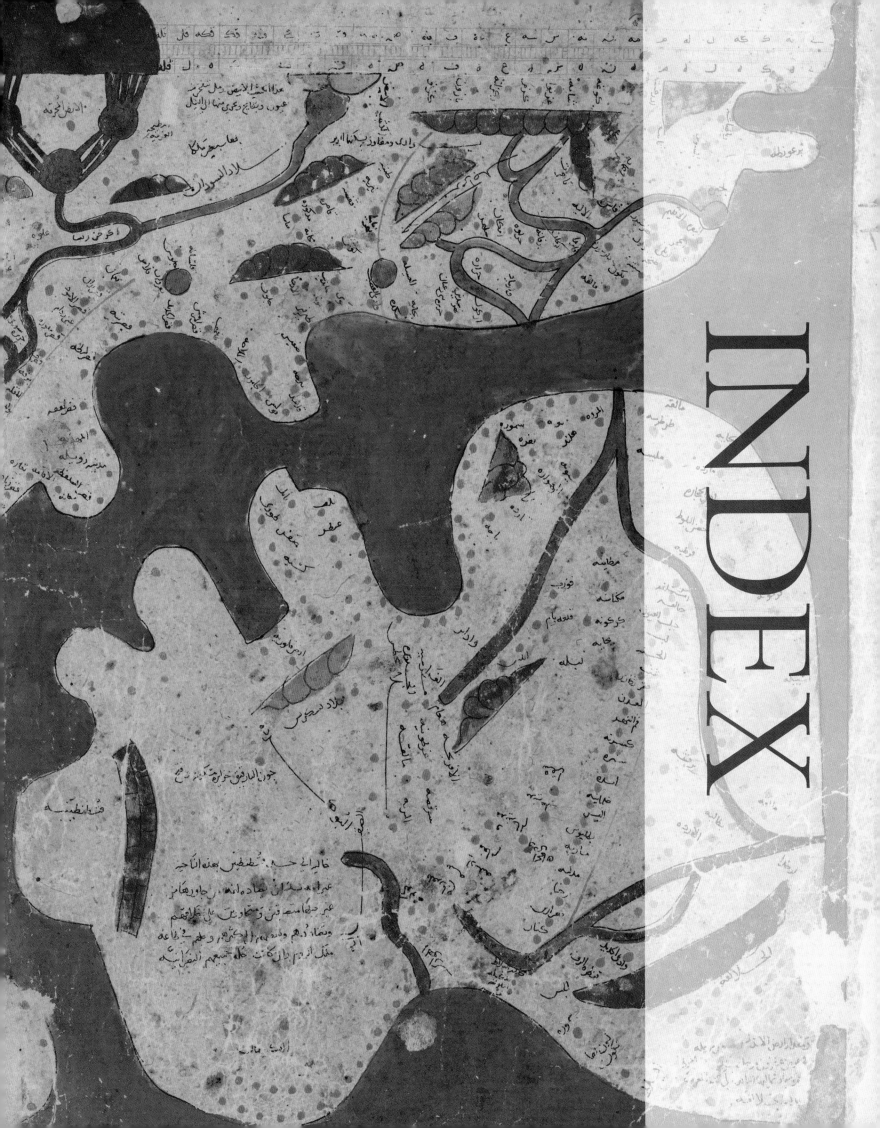

INDEX

Index

A

Afghanistan, *Mappa* (Boetti) 232-35
Africa
  Cape of Good Hope 74
  Cartogram (Worldmapper) 239
  as circumnavigable, Fra Mauro world
    map 73, 74, 75
  as circumnavigable, *Kangnido* map
    (Kwŏn Kŭn) 67
  Dymaxion Map (Fuller) 223
  East African drift system, *World Ocean
    Floor* (Tharp and Heezen) 231
  Equal Area World Map (Peters) 228
  Fra Mauro world map 72-75
  first map of America (Waldseemüller)
    12-13, 86-89
  "four-by-four" grids, *Carte Pisane*
    54-55
  Hereford Mappa Mundi, mythical places
    and monsters 58, 92
  Inhabited Quarter (Isfahani) 140
  Livingstone's map of Africa
    15, 198-201
  Mappa (Boetti) 234
  Missionary Map (Miller) 202-03
  nautical chart (Zheng He) 134, 136-37
  Piri Re'is Map 93
  Ptolemy's world map 24
  Southern, *Ten Thousand Countries of
    the Earth* (Ricci, Li and Zhang) 129
  trans-Saharan trade routes,
    Catalan Atlas 64
  West African coast, Juan de la Cosa's
    world chart 79
Alexander the Great
  Catalan Atlas 64
  Indian world map 176, 179
  Piri Re'is Map 90
Alexander's Barrier, *Book of Curiosities* 43
Alexandria
  Kangnido map (Kwŏn Kŭn) 67
  Peutinger Map 31
  Royal Library of 24
alphabetical table, *New Map of France*
    (Cassini de Thury) 164
America
  British Colonies in North America
    (Mitchell) 172-75

California as island, *New Map of the
    World* (Blaeu) 145
California, land passage to (Kino)
    160-61
Cartogram (Worldmapper) 239
first map of (Waldseemüller)
    12-13, 86-89
"Indian Territory" map (Tanner)
    15, 190-91
New American Atlas (Tanner) 191
New England Map (Foster) 150-53
portolan sailing chart (Pizzigano)
    68, 70
*Slave Population of the Southern States
    of the US* (Hergesheimer) 194-97
Ten Thousand Countries of the Earth
    (Ricci) 128
see also Canada; South America
Antarctica
  Dymaxion Map (Fuller) 222-23
  Equal Area World Map (Peters) 229
  International Map of the World
    (Penck) 215
  *A New and Enlarged Description
    of the Earth* (Mercator) 113
  Piri Re'is Map 93
Arabian Peninsula
  Book of Curiosities 40-43
  Inhabited Quarter (Isfahani) 140
  nautical chart (Zheng He) 134
Aristotle, climate divisions 10, 43
Arte Povera ("Poor Art") (Boetti) 233, 235
Asia
  Asia Minor, *Carte Pisane* 55
  Juan de la Cosa's world chart 79
  Map of All Under Heaven 180-83
  Selden Map 130-33
  Silk Route and Marco Polo 63,
    64-65
  Southeast, Cartogram (Worldmapper)
    238
  see also individual countries
astrology
  Catalan Atlas 65
  Ptolemy 24, 27
astronomy
  Book of Curiosities 11, 40-43
  British Colonies in North America
    (Mitchell) 174

Catalan Atlas 65
Corrected Map of France (Picard and
    La Hire) 155
Dunhuang Star Chart 36-39
Fra Mauro world map 72-75
Jain cosmological map 168-71
nautical chart (Zheng He) 136, 137
New Map of France (Cassini de
    Thury) 164
New Map of the World (Blaeu)
    144-45
Ten Thousand Countries of the Earth
    (Ricci) 128
uranomancy 36
Atlante Nautico (Bianco) 71
Atlantic Ocean
  portolan sailing chart (Pizzigano)
    68-71
  World Ocean Floor (Tharp and
    Heezen) 230
atlas
  first use of term (Mercator) 110
  world's first, *Theatrum Orbis Terrarum*
    (Ortelius) 159
*Atlas of the Counties of England and
    Wales* (Saxton) 116
Atlas Maior (Blaeu) 143
Augsburg map (Seld) 96-99
Australia
  Dymaxion Map (Fuller) 223
  International Map of the World
    (Penck) 215
  New Map of the Whole World
    (Blaeu) 143, 144
Aztecs
  Codex Florentine 105
  Codex Mendoza 105, 130
  Tenochtitlan map 13, 104-05
  see also South America

B

Babylonian world map 22-23
Bache, Alexander, *Slave Population of
    the Southern States of the US*
    (Hergesheimer) 15, 194-97
Bar-Jacob, Abraham, Holy Land map
    14, 156-59

Beatus world map 59
Beck, Harry
London Underground map 216–19
Paris Metro 216
Bedolina petroglyph 20–21
Bering Strait, *International Map of the World* (Penck) 215
Bianco, Andrea, *Atlante Nautico* 71
Black Sea
Carte Pisane 53, 55
Entertainment for He Who Longs to Travel the World (al-Idrīsī) 48
Blaeu, Joan
Atlas Maior 143
New Map of the Whole World 142–45
Boetti, Alighiero
Arte Povera ("Poor Art") 233
Mappa 16, 232–35
Book of Curiosities 11, 40–43
Booth, Charles, *London Poverty* 15, 197, 204–20
Brahe, Tycho 143
Brasil (legendary), portolan sailing chart (Pizzigano) 71
Brazil, Juan de la Cosa's world chart 76, 78
Britannia Atlas Road Map (Ogilby) 146–49
British Colonies in North America (Mitchell) 172–75
British Isles
Carte Pisane 53, 55
Cartogram (Worldmapper) 239
Greenwich 214
Hereford Mappa Mundi 58
Ireland, Down Survey 149
London *see* London
Northumbria county map 116–17
Nova Utopia (Walter) 241
portolan sailing chart (Pizzigano) 71
Ptolemy's world map 26
geological map (Smith) 15, 184–87
Ten Thousand Countries of the Earth (Ricci) 128
Brongniart, Alexandre, "geognostic" map of Paris Basin 187
Buddhism, *Map of All Under Heaven* 181, 182
Burma, nautical chart (Zheng He) 137
Byzantines, Madaba mosaic map 34–37

C

Cabot, John 76, 109
Canada
New France (Gastaldi) 106–09

Ten Thousand Countries of the Earth (Ricci) 128
see also America
Canary Islands
Catalan Atlas 64
Entertainment for He Who Longs to Travel the World (Al-Idrīsī) 48
Inhabited Quarter (Isfahani) 140
longitude starting from 140
Cape Verde, portolan sailing chart 71
Caribbean Islands, Juan de la Cosa's world chart 76, 78
Carte Pisane 11, 52–55
Carthage, *Peutinger Map* 30
Cartogram (Worldmapper) 16, 236–39
Cary, John, geological map 184
Caspian Sea, *Inhabited Quarter* (Isfahani) 141
Cassini de Thury, César-François, *New Map of France* 14, 162–65
Catalan Atlas 12, 62–65
reproduction, Juan de la Cosa's world chart 79
China
Map of All Under Heaven 180–83
compass use 132, 133
Dunhuang Star Chart 36–39
Fra Mauro world map 74
Linde calendar 37
Map of the Tracks of Yu 44–45
Ming Dynasty, *Kangnido* map (Kwŏn Kŭn) 66–67
Ming Dynasty, *Selden Map* 130–33
nautical chart (Zheng He) 134–37
Ten Thousand Countries of the Earth (Ricci) 126–29
cholera, *Geographical and Statistical Account of the Epidemic Cholera* (Tanner) 191
cholera map (Snow) 15, 192–93
Map of All Under Heaven 180–83
Christianity
Beatus world map 59
Catalan Atlas 62–65
Fra Mauro world map 72–75
Hereford Mappa Mundi 56–59
Holy Land *see* Holy Land
influence, *Peutinger Map* 31
Jesuit missionary work, *Ten Thousand Countries of the Earth* (Ricci) 126–29
Madaba mosaic map 34–37
Missionary Map (Miller) 203
Psalter World Map 50
Sawley Map 50–51
Vatican gallery of maps (Danti) 118–21

climate divisions, *Book of Curiosities* 43
"clip-mapping", Google Earth 245
cloud divination, *Dunhuang Star Chart* 38–39
Codex Mendoza, Aztecs 105, 130
Columbus, Christopher 12
and first map of America (Waldseemüller) 86, 88
Hispaniola discovery 92
Juan de la Cosa's world chart 76–79
Piri Re'is Map 90
compass rose
Britannia Atlas Road Map (Ogilby) 147, 149
Juan de la Cosa's world chart 79
New England map (Foster) 152–53
Piri Re'is Map 90, 93
Selden Map 130, 132
compass use
China 132, 133
Hereford Mappa Mundi 58
Constantinople (Istanbul)
Book of Curiosities 41
Indian world map 176
Kangnido map (Kwŏn Kŭn) 67
Peutinger Map 31
Copernicus 143, 225
Corrected Map of France (Picard and La Hire) 154–55
cosmology *see* astronomy
Cresques, Abraham, *Catalan Atlas* 62–65
cuneiform text 22–23
Cuvier, Georges, "geognostic" map of Paris Basin 187

D

Da Vinci, Leonardo, map of Imola 13, 84–85
Danti, Egnazio
Guardaroba, Florence 119
Vatican gallery of maps 118–21
Darwin, Charles, fossil deposits 187
De' Barbari, Jacopo, Venice map 80–83, 97, 99
De Bry, Theodor, colonial Virginia map 175
De la Cosa, Juan, world chart 12, 76–79
Dead Sea, Madaba mosaic map 37
Delos (island), *Sawley Map* 50–51
directional orientation, lack of, *Catalan Atlas* 62–65
Dorling, Danny 16, 237
Duchamp, Marcel 233
Dunhuang Star Chart 10, 36–39

Dutch East India Company 122, 124, 125, 143
Dymaxion Map (Fuller) 16, 220-23

E

Earth's circumference measurement, first 10
east at top of map, *Hereford Mappa Mundi* 56-59
Egypt
   Alexandria *see* Alexandria
   Peutinger Map 31
   Sawley Map 50
embroidery, *Mappa* (Boetti) 232-35
Emslie, John, geological world map 187
engraving, *A New and Enlarged Description of the Earth* (Mercator) 110-13
Entertainment for He Who Longs to Travel the World (Al-Idrīsī) 46-49
Equal Area World Map (Peters) 16, 226-29
Eratosthenes, *Geography* 10
Europe
   Dymaxion Map (Fuller) 222
   Equal Area World Map (Peters) 229
   Indian world map 179
   International Map of the World (Penck) 214-15
   Kangnido map (Kwŏn Kŭn) 66-67
   Mappa (Boetti) 235
   Middle Ages, *Hereford Mappa Mundi* 56-59
   Ten Thousand Countries of the Earth (Ricci) 128
   as wasteland, *Map of All Under Heaven* 183
   see also individual countries

F

Ferrer, Jacme (Jaume), *Catalan Atlas* 64
fieldwork techniques, *Corrected Map of France* (Picard and La Hire) 154-55
Fischer, Father Joseph, first map of America, discovery of 89
Foster, John, New England map 14, 150-53
"four-by-four" grids, *Carte Pisane* 54
Fra Mauro world map 72-75

France
   Corrected Map of France (Picard and La Hire) 154-55
   "Dieppe School" of mapmaking 109
   New Map of France (Cassini de Thury) 14, 162-65
   Paris *see* Paris
Fugger, Jakob, map of Augsburg (Seld) 97, 98
Fuller, Buckminster, *Dymaxion Map* 16, 220-23

G

Galileo, lunar landscape 225
Gall, Reverend James, orthographic equal-area projection 229
Gastaldi, Giacomo, *New France* (Canada) 106-09
geocentric system (Ptolemy) 128, 144
"geognostic" map of Paris Basin (Cuvier and Brongniart) 187
geological mapmaking 184-87
   geological world map (Emslie) 187
"George III's map", *British Colonies in North America* (Mitchell) 172-75
"geospatial" applications, Google Earth 17, 244-45
Germany
   Kangnido map (Kwŏn Kŭn) 67
   map of Augsburg (Seld) 96-99
global security, Google Earth 245
global warming, *Dymaxion Map* (Fuller) 222
Gog and Magog (monsters)
   Alexander's Barrier 41, 43
   Catalan Atlas 64
   Indian world map 179
   Juan de la Cosa's world chart 79
   Sawley Map 50
   see also monsters
Google Earth 17, 244-45
gores (curved segments) for creating globes 89
Graham, Henry, *Slave Population of the Southern States of the US* (Hergesheimer) 15, 194-97
graticule (grid of coordinates) 24
   Inhabited Quarter (Isfahani) 138-41
   Ten Thousand Countries of the Earth (Ricci) 127
   see also latitude and longitude
Greeks *see* Aristotle; Ptolemy, Claudius

Greenland, *Equal Area World Map* (Peters) 228
Greenwich 214
Guerry, André-Michel, chloropleth map of France 197

H

Haggadah (Jewish prayer book) illustration, Holy Land map (Bar-Jacob) 156-59
health *see* cholera
Hebrew, Holy Land map (Bar-Jacob) 156-59
Heezen, Bruce, *World Ocean Floor* 16, 17, 230-31
heliocentric theories 143, 225
Hereford Mappa Mundi 11, 56-59
Hergesheimer, Edwin, *Slave Population of the Southern States of the US* 15, 194-97
Holbein, Ambrosius
   Utopia, map of 13, 94-95, 243
Holy Land
   Carte Pisane 53
   map (Bar-Jacob) 14, 156-59
   Sawley Map 50-51
   see also Christianity
Holy Roman Empire 96, 99
Hubbard, William, *Narrative of the Troubles with the Indians in New-England* (book) 151

I

ichnographic plan, map of Imola (Da Vinci) 85
Al-Idrīsī, Al-Sharif, *Entertainment for He Who Longs to Travel the World* 11, 46-49
imaginary elements, *Map of All Under Heaven* 181, 182, 183
Imola, map of (Da Vinci) 13, 84-85
"Impossible Black Tulip" of cartography, *Ten Thousand Countries of the Earth* (Ricci) 126-29
India
   Cartogram (Worldmapper) 238
   Indian world map 176-79
   Inhabited Quarter (Isfahani) 138-41
   Jain cosmological map 170
   Juan de la Cosa's world chart 79
   Kangnido map (Kwŏn Kŭn) 66-67
   nautical chart (Zheng He) 136

Selden Map 132
Indian Ocean, *Universal Chart*
    (Ribeiro) 103
"Indian Territory" map (Tanner) 15,
    190-91
Indonesia
    Equal Area World Map (Peters) 229
    in first map of America (Waldseemüller)
        89
    International Map of the World
        (Penck) 215
    Java *see* Java
    Molucca Islands *see* Molucca Islands
    Sumatra *see* Sumatra
    Universal Chart (Ribeiro) 103
Inhabited Quarter (Isfahani) 14, 138-41
International Map of the World
    (Penck) 16, 212-15
Iraq, Babylonian world map 22-23
Ireland, Down Survey 149
Isfahani, Sadiq, *Inhabited Quarter*
    14, 138-41
Islam
    Book of Curiosities 11, 40-43
    Indian world map 176-79
    world maps, *Entertainment for He
        Who Longs to Travel the World*
        (Al-Idrīsī) 46-49
Island of the Jewel (al-Khwarizmi) 43
Istanbul *see* Constantinople
Italy
    Bedolina petroglyph 20-21
    Florence, *Guardaroba* (Danti) 119
    Kangnido map (Kwŏn Kŭn) 67
    map of Imola (Da Vinci) 84-85
    Rome *see* Rome
    Valcamonica petroglyphs 20
    Venice *see* Venice
    Venice map 80-83

J

Jainism, Jain cosmological map 168-71
Japan
    Map of All under Heaven 182
    Fra Mauro world map 73, 75
    Hokkaido to Kyushu (Tadataka)
        15, 188-89
    Kangnido map (Kwŏn Kŭn) 66-67
    Piri Re'is Map 92
    Ten Thousand Countries of the Earth
        (Ricci) 129
Java
    in first map of America
        (Waldseemüller) 89

Fra Mauro world map 75
Molucca Islands map (Plancius)
    124-25
Universal Chart (Ribeiro) 103
see also Indonesia
Jerusalem
    Catalan Atlas 62
    Fra Mauro world map 73
    Hereford Mappa Mundi 57, 58
    Juan de la Cosa's world chart 79
    Madaba mosaic map 34, 36-37
Jordan, Madaba mosaic map 34-37
Judaism, Holy Land map (Bar-Jacob)
    156-59

K

Kangnido map (Kwŏn Kŭn) 12, 66-67
al-Khwarizmi, Island of the Jewel 43
Kino, Eusebio Francisco, land passage to
    California 14, 160-61
klimata (climes) 43, 49
Kolb, Anton (publisher), Venice map
    81
Korea
    Kangnido map (Kwŏn Kŭn) 66-67
    Map of all Under Heaven 180-83
    Ten Thousand Countries of the Earth
        (Ricci) 129
Kublai Khan, *Catalan Atlas* 64
Kwŏn Kŭn, *Kangnido* map 12, 66-67

L

La Hire, Philippe, *Corrected Map
    of France* 154-55
latitude and longitude
    graticule *see* graticule
    Greenwich 214
    klimata (Ptolemy) 49, 138,
        141, 222
    longitude starting from Canary
        Islands 140
    Mercator's projection *see* Mercator,
        Gerard
    meridian line, Juan de la Cosa's
        world chart 77
    modified conic, *International Map
        of the World* (Penck) 213
    New Map of France (Cassini de
        Thury) 164
Li Chunfeng
    Dunhuang Star Chart 36-39
    Linde calendar 37

Li Zhizao, *Ten Thousand Countries of
    the Earth* 14, 126-29
Lincoln, Abraham, *Slave Population of
    the Southern States of the US*
    (Hergesheimer) 195
Livingstone, David, map of Africa
    15, 198-201
London
    Britannia Atlas Road Map (Ogilby) 148
    Carte Pisane 55
    cholera map (Snow) 192-93
    poverty map (Booth) 15, 204-07
    Subterranea (Walter) 241
    Underground map (Beck) 216-19
    see also British Isles
London poverty map (Booth) 15, 204-07
Lunar landings map (NASA) 17, 224-25

M

Madaba mosaic map 10, 34-37
Madagascar, *Inhabited Quarter* (Isfahani)
    140, 141
Magellan, Ferdinand, global
    circumnavigation 100-01, 113
Maldives, nautical chart (Zheng He)
    134, 136
Mao Yuanyi, *Wubei Zhi* 134
map, definition 7
Map of All Under Heaven 180-83
Mappa (Boetti) 16, 232-35
mappa mundi *see* world maps,
    mappa mundi
Marco Polo
    Catalan Atlas 63, 64-65
    Siam 125
Marshall Islands stick chart 208-09
mathematical calculations
    Eratosthenes 10
    first map of America (Waldseemüller)
        89
    *A New and Enlarged Description of
        the Earth* (Mercator) 110-13
    Venice map 83
    see also statistics
Mecca
    Book of Curiosities 41, 43
    *Entertainment for He Who Longs to
        Travel the World* (Al-Idrīsī) 47
    Indian world map 176, 178
Mediterranean
    Book of Curiosities 43
    Carte Pisane 52-55
    as centre of inhabited world 26
    Kangnido map (Kwŏn Kŭn) 67

Ptolemy's world map and over-
    estimation of size 24-27
Mercator, Gerard and Mercator's
    projection
    Antarctica 222
    criticism of, *Equal Area World Map*
        (Peters) 226, 228, 229
    Molucca Islands (Plancius) 122-25
    Ten Thousand Countries of the Earth
        (Ricci) 127
    terrestrial globe 113
    *A New and Enlarged Description of
        the Earth* 12, 110-13
    see also latitude and longitude
Mexico
    Aztecs *see* Aztecs
    Mappa (Boetti) 234
Middle East, *Mappa* (Boetti) 235
military maps
    International Map of the World
        (Penck) 215
    see also political maps
Miller, William, *Missionary Map* 202-03
Millionth Map of the World
    (*International Map of the World*)
    (Penck) 212-15
Missionary Map (Miller) 202-03
Mississippi River
    British Colonies in North America
        (Mitchell) 175
    *Slave Population of the Southern States
        of the US* (Hergesheimer) 196-97
Mitchell, John, *British Colonies in North
    America* 172-75
Molucca Islands
    Plancius 122-25
    Selden Map 133
    Universal Chart (Ribeiro) 12-13,
        100-03
    see also Indonesia
monsters
    Gog and Magog *see* Gog and Magog
    Indian world map 176, 179
    *A New and Enlarged Description of the
        Earth* (Mercator) 112
    Molucca Islands map (Plancius) 124
    New France (Gastaldi) 108
    portolan sailing chart (Pizzigano) 68-71
    Ten Thousand Countries of the Earth
        (Ricci) 129
"mopeds and motorcycles" cartogram
    239
More, Thomas, *Utopia* 13, 94-95,
    241, 243
Morocco
    Book of Curiosities 43

Carte Pisane 55
mosaic, Madaba mosaic map 37
Mozambique, nautical chart
    (Zheng He) 137
Mughal culture, *Inhabited Quarter*
    (Isfahani) 138-41
Muslim world *see* Islam

N

NASA, Lunar landings map 17, 224-25
navigational maps
    Marshall Islands stick chart 208-09
    *A New and Enlarged Description of the
        Earth* (Mercator) 110-13
    Selden Map 130-33
    see also portolan sailing charts
Netherlands, *New Map of the Whole
    World* (Blaeu) 142-45
New American Atlas (Tanner) 191
New England map (Foster) 14, 150-53
New France (Gastaldi) 106-09
New Map of France (Cassini de Thury)
    162-65
New Map of the Whole World (Blaeu)
    142-45
New Universal Atlas (Tanner) 191
New York, *New France* (Gastaldi) 108
Newfoundland
    Juan de la Cosa's world chart 767
    New France (Gastaldi) 106-09
Nile River Delta
    Indian world map 178
    Peutinger Map 31
Nile River source
    Book of Curiosities 42-43
    *Entertainment for He Who Longs to
        Travel the World* (Al-Idrīsī) 48
    Livingstone's map of Africa 198-201
North Pole
    Dunhuang Star Chart 38
    Dymaxion Map (Fuller) 222
    International Map of the World
        (Penck) 214
    *A New and Enlarged Description of
        the Earth* (Mercator) 112, 113
Nova Utopia, (Walter) 16, 240-43

O

ocean floor, *World Ocean Floor* (Tharp
    and Heezen) 16, 17, 230-31
Ogilby, John, *Britannia Atlas
    Road Map* 146-49

Oliva, Joan, portolan chart 55
Orbis Terrarum miltary map, Roman
    Empire 31
Ordnance Survey
    International Map of the World
        (Penck) 215
    military maps 215
    origins 117
    standard scale adoption 147
Ortelius, Abraham, *Theatrum Orbis
    Terrarum* (world's first atlas)
    127, 159
orthographic equal-area projection,
    Equal Area World Map (Peters)
    229
Ottoman Empire
    Piri Re'is Map 90
    Venice historical chart 93
oval projection, *Ten Thousand Countries
    of the Earth* (Ricci) 126-29

P

Pacific drift, *World Ocean Floor*
    (Tharp and Heezen) 231
Pacioli, Luca, map of Venice (Seld) 83
Paris
    "geognostic" map of Paris Basin
        (Cuvier and Brongniart) 187
    meridian, *Corrected Map of France*
        (Picard and La Hire) 155
    meridian, *New Map of France*
        (Cassini de Thury) 164
    Metro (Beck) 216
    see also France
Penck, Albrecht, *International Map
    of the World* 16, 212-15
Pergamon, *Peutinger Map* 31
Periodos Ges ("Circuit of the Earth") 10
Peters, Arno, *Equal Area World Map*
    16, 226-29
petroglyph, Bedolina 20-21
Peutinger Map 28-31
Pharos *see* Alexandria
"physiographic" maps, *World Ocean Floor*
    230-31
Picard, Jean, *Corrected Map of France*
    154-55
Piri Re'is Map 90-93
Pizzigano, Zuane, portolan sailing chart
    12, 68-71
Plancius, Petrus
    Molucca Islands map 122-25
    Ten Thousand Countries of the Earth
        (Ricci) 129

political maps
  International Map of the World
    (Penck) 16, 212-15
  Teixeira, Domingos, world map 103
  Universal Chart (Ribeiro) 100-03
  see also military maps
portolan sailing charts 12-13, 75
  Carte Pisane 11, 52-55
  Catalan Atlas 12, 62-65
  Molucca Islands map (Plancius)
    122-25
  nautical chart (Zheng He) 134-37
  Piri Re'is Map 90-93
  Pizzigano, Zuane 12, 68-71
  see also navigational maps
poverty map 204-07
prehistoric cartographers, petroglyph
  20-21
projection techniques
  Mercator's projection *see* Mercator,
    Gerard
  orthographic equal-area projection
    229
  oval projection 126-29
  Ptolemy's world map 24, 27
Psalter World Map 50
Ptolemy, Claudius
  ecumene (inhabited world) 24, 26,
    88, 138
  geocentric system 128, 144
  Geography (book) 10, 24, 106,
    176, 179
  influence of 41, 87, 88, 89, 110
  klimata (climes) 43, 49
  and *Piri Re'is Map* 90
  world map 24-27, 48, 49

R

radial lines, map of Imola
  (Da Vinci) 85
Ramusio, Giovanni Battista, *Navigations
  and Voyages* 106
Rassam, Hormuzd, Babylonian World
  Map 22
Ratzel, Friedrich, *Lebensraum* (living
  space) concept 213
reference chart, London Underground
  map (Beck) 218
relief depiction, *Britannia Atlas Road Map*
  (Ogilbe) 147, 148
religion *see* Buddhism; Christianity; Islam;
  Jainism; Judaism
rhumb lines, portolan sailing chart
  (Pizzigano) 68

Ribeiro, Diogo, *Universal Chart* 12-13,
  100-03
Ricci, Matteo, *Ten Thousand Countries
  of the Earth* 14, 126-29
Riccioli, Giovanni Battista, lunar
  landscape 225
Richard of Haldingham, *Hereford Mappa
  Mundi* 56-59
Roman Empire
  map of Augsburg (Seld) 98-99
  Orbis Terrarum military map 31
  Peutinger Map 28-31
Rome
  Peutinger Map 30
  Vatican gallery of maps (Danti)
    118-21
  see also Italy
Royal Geographical Society (RGS),
  military maps 215
Russia
  Cartogram (Worldmapper) 238
  Mappa (Boetti) 235

S

Sahara Desert, *Kangnido* map
  (Kwŏn Kŭn) 67
St Christopher, Juan de la Cosa's
  world chart 78
Sardinia, *Carte Pisane* 54
Satanazes (Isle of Devils), portolan
  sailing chart (Pizzigano) 70
Sawley Map 50-51
Saxton, Christopher
  *Atlas of the Counties of England
    and Wales* 116
  Northumbria, map of 116-17
scale
  first cartographic grid showing, *Map
    of the Tracks of Yu* 44-45
  of shade, *Slave Population of the
    Southern States of the US*
    (Hergesheimer) 196
scale bar
  Carte Pisane 54
  Island of the Jewel 43
  New England map (Foster)
    152
  Northumbria map 117
  Piri Re'is Map 90, 93
  portolan sailing chart (Pizzigano)
    68, 70
  Selden Map 130, 132
Scythia, cannibalism, *Hereford
  Mappa Mundi* 59

Seld, Jörg, map of Augsburg 96-99
Selden Map 130-33
Sicily, *Entertainment for He Who Longs to
  Travel the World* (Al-Idrīsī) 46-49
*Slave Population of the Southern States
  of the US* (Hergesheimer) 15, 194-97
Smith, William, geological map, Britain
  15, 184-87
Snow, John, cholera map 15, 192-93
social equality, *Equal Area World Map*
  (Peters) 226-29
social mapping, *London Poverty*
  (Booth) 204-07
solar system, heliocentric theory, *New
  Map of the Whole World* (Blaeu)
  142, 143
South America
  Aztecs *see* Aztecs
  Cartogram (Worldmapper) 239
  Dymaxion Map (Fuller) 223
  Equal Area World Map (Peters) 228
  International Map of the World
    (Penck) 215
  Juan de la Cosa's world chart 76-79
  Mappa (Boetti) 234
  *A New and Enlarged Description of
    the Earth* (Mercator) 113
  Mexico *see* Mexico
  monstrous creatures, *Piri Re'is
    Map* 92
  Piri Re'is Map 90-93
  see also America
south depicted at top of map
  Book of Curiosities 40-43
  *Entertainment for He Who Longs to
    Travel the World* (Al-Idrīsī) 46-49
  Fra Mauro's world map 72-75
southeast depicted at top of map, Holy
  Land map (Bar-Jacob) 156-59
Spain
  Book of Curiosities 41, 43
  Tordesillas meridian line 72
  Universal Chart (Ribeiro) 100-03
spherical Earth on flat surface
  Dymaxion Map (Fuller) 220-23
  first depiction, *A New and Enlarged
    Description of the Earth* (Mercator)
    110-13
"Spring of Life", Indian world map 178
Sri Lanka (Taprobana)
  Catalan Atlas 65
  Inhabited Quarter (Isfahani) 140, 141
  nautical chart (Zheng He) 136
  Ptolemy's world map 27
statistics
  Cartogram (Worldmapper) 236-39

chloropleth mapping technique,
Slave Population of the Southern States of the US (Hergesheimer) 194–97
cholera map (Snow) 193
see also mathematical calculations
Stein, Aurel, *Dunhuang Star Chart* 36
stele carving, *Map of the Tracks of Yu* 44–45
stick charts, Marshall Islands stick chart 208–09
strata delineation, Britain (Smith) 184–87
Subterranea (Walter) 241
Sumatra
Inhabited Quarter (Isfahani) 140
nautical chart (Zheng He) 137
Selden Map 132
Universal Chart (Ribeiro) 103
see also Indonesia
survey data
London Poverty (Booth) 207
theodolite use *see* theodolite use
topographic surveys, map of Augsburg (Seld) 99
symbol standardization
Britannia Atlas Road Map (Ogilby) 147, 148, 149
New Map of France (Cassini de Thury) 163
Ordnance Survey *see* Ordnance Survey

T

Tadataka, Inō, *Japan, Hokkaido to Kyushu* 15, 188–89
Tanner, Henry Schenck, "Indian Territory" map 15, 190–91
Taprobana *see* Sri Lanka
Tasmania, *New Map of the Whole World* (Blaeu) 143, 144, 145
Teixeira, Domingos, political map 103
temperature zones, *Dymaxion Map* (Fuller) 220–23
Ten Thousand Countries of the Earth (Ricci) 14, 126–29
Tenochtitlan, Aztec map 13, 104–05
Tharp, Marie, *World Ocean Floor* 16, 17, 230–31
Theatrum Orbis Terrarum (world's first atlas) (Ortelius) 159
theodolite use
Down Survey 149
map of Imola (Da Vinci) 85
see also survey data

treasure ships, nautical chart (Zheng He) 137
triangulation methods
Britannia Atlas Road Map (Ogilby) 147
Corrected Map of France (Picard and La Hire) 154–55
New Map of France (Cassini de Thury) 163, 164, 165
Tropic of Cancer
Juan de la Cosa's world chart 76–77
Piri Re'is Map 93

U

Universal Chart (Ribeiro) 12–13, 100–03
USA *see* America
Utopia (book) 13, 94–95, 241, 243
Utopia, map of (Holbien) 13, 94–95, 243

V

Van Diemen's Land *see* Tasmania
Vatican gallery of maps (Danti) 118–21
Venice
as centre of medieval and Renaissance mapmaking 75
historical chart, Ottoman Empire 93
map (De' Barbari) 80–83, 97, 99
see also Italy
Verrazzano, Giovanni da, North American exploration 106, 108
Vespucci, Amerigo, first map of America (Waldseemüller) 86, 88

W

Waldseemüller, Martin
first map of America 12–13, 86–89
gores (curved segments) for creating globes 89
Walter, Stephen
Nova Utopia 13, 16, 240–43
Subterranea 241
Waqwaq Islands, *Inhabited Quarter* (Isfahani) 140
Weiditz, Hans, map of Augsburg (Seld) 97
Wesel, Moses, *Amsterdam Haggadah* 157
west depicted at top of map, *New England Map* (Foster) 150–53
wind heads, Venice map 83

wind rose, Juan de la Cosa's world chart 79
world maps
Babylonian world map 22–23
Entertainment for He Who Longs to Travel the World (Al-Idrīsī) 46–49
Equal Area World Map (Peters) 16, 226–29
geological world map (Emslie) 187
Indian world map 176–79
International Map of the World (Penck) 16, 212–15
A New and Enlarged Description of the Earth (Mercator) 12, 110–13
New Map of the Whole World (Blaeu) 142–45
Ptolemy's world map 24–27, 48, 49
Theatrum Orbis Terrarum (world's first atlas) (Ortelius) 159
World Ocean Floor (Tharp and Heezen) 16, 17, 230–31
world maps, mappa mundi 10
Beatus world map 59
Catalan Atlas 62–65
Map of All Under Heaven, stylistic links 181
Fra Mauro world map 72–75
Hereford Mappa Mundi 11, 56–59
Juan de la Cosa's world chart 12, 76–79
Psalter World Map 50
Sawley Map 50–51
Worldmapper, Cartogram 236–39
Wubei Zhi (Mao Yuanyi) 134

Y

Yu ji tu (*Map of the Tracks of Yu*) 10–11, 44–45

Z

Zanzibar 200–01
Zhang Wentao, *Ten Thousand Countries of the Earth* 14, 126–29
Zheng He
Cape of Good Hope 74
nautical chart 134–37

Acknowledgments

Dorling Kindersley would like to thank the following people for their assistance with this book:
Smithsonian reviewer, Jim Harle, map curator volunteer, National Museum of Natural History; Richard Gilbert, Georgina Palffy, Priyaneet Singh, and Phil Wilkinson for editorial assistance; Margaret McCormack for compiling the index; Martin Copeland for picture research assistance; Vishal Bhatia for DTP assistance; and Carrie Mangan for admin assistance.

The publisher would specially like to thank the following people for their help in providing DK with images:
Luca Giarelli for the Bedolina Petroglyph, Mark Pretlove for the Britannia Road Atlas, and Patrick Rodgers for the Haggadah.

The publisher would like to thank the following for their kind permission to reproduce their photographs:

(Key: a-above; b-below/bottom; c-centre; f-far; l-left; r-right; t-top)

1 The Bodleian Library, University of Oxford: MS. Pococke 375, fol. 3b-4a (cl). **Getty Images:** British Library / Robana (cr). **Map Reproduction Courtesy of the Norman B. Leventhal Map Center at the Boston Public Library:** (c). **2-3 Museo Naval de Madrid. 4-5 The Library of Congress, Washington DC:** G3200 1507 .W3 Vault. **6-7 Corbis:** (t). **Library of the London School of Economics & Political Science:** Booth's Maps Descriptive of London Poverty 1898-1899 (b). **State Library of New South Wales:** Safe / M2 470 / 1617 / 1 (c). **8-9 Getty Images:** A. Dagli Orti / De Agostini. **10 © 2014, Biblioteca Apostolica Vaticana:** Urb.gr.82.ff.60v-61r by permission of Biblioteca Apostolica Vaticana, with all rights reserved (b). **11 The Bodleian Library, University of Oxford:** MS Arab. c. 90, fols. 23b-24a (b). **The Mappa Mundi Trust and Dean and Chapter of Hereford Cathedral:** (tr). **12-13 The Library of Congress, Washington DC:** G3200 1507 .W3 Vault (bc). **12 Museo Naval de Madrid:** (bl). **13 The Bodleian Library, University of Oxford:** MS. Arch. Selden. A1, fol. 2r (br). **14 The Rosenbach of the Free Library of Philadelphia Foundation:** [Haggadah]. Seder Hagadah shel Pesah: ke-minhag Ashkenaz u-Sefarad. Amsterdam: Bevet ha-meshutafim…Shalit, [5] 455 [i.e. 1695] (b). **15 Library of the London School of Economics & Political Science:** Booth's Maps Descriptive of London Poverty 1898-1899 (bl). **16 Copyright by Marie Tharp 1977/2003. Reproduced by permission of Marie Tharp:** (br). **Map Reproduction Courtesy of the Norman B. Leventhal Map Center at the Boston Public Library:** (bl). **17 Google Earth:** Map data: Google, DigitalGlobe, SIO, NOAA, U.S. Navy, NGA, GEBCO (bl). **U.S. Air Force:** (br). **18-19 The Bodleian Library, University of Oxford:** MS. Pococke 375, fol. 3b-4a. **20-21 Luca Giarelli. 21 Alamy Images:** LOOK Die Bildagentur der Fotografen GmbH (cra). **22-23 Corbis. 22 Dorling Kindersley:** (bl). **24-27 © 2014, Biblioteca Apostolica Vaticana:** Urb.gr.82.ff.60v-61r by permission of Biblioteca Apostolica Vaticana, with all rights reserved. **24 Getty Images:** The Bridgeman Art Library (bc). **27 Alamy Images:** AF Fotografie (cra). **28-31 ÖNB/Wien:** Cod. 324. **29 Stadt Augsburg:** (cra). **31 Alamy Images:** Universal Art Archive (br). **32-33 akg-images:** Erich Lessing. **32 akg-images:** (tr). **Alamy Images:** MB Travel (bl). **34 akg-images:** (tl); Erich Lessing (bc). **Corbis:** The Gallery Collection (cl). **34-35 akg-images:** Erich Lessing (c). **35 akg-images:** Erich Lessing (cr, tr). **Alamy Images:** Art Directors & TRIP (bl). **Rough Guides:** Jean-Christophe Godet (br). **36-39 akg-images:** British Library. **39 TopFoto.co.uk:** World History Archive (tr). **40-43 The Bodleian Library, University of Oxford:** MS Arab. c. 90, fols. 23b-24a. **43 The Art Archive:** Bodleian Libraries, The University of Oxford (cra). **44-45 The Library of Congress, Washington DC:** G7821.C3 1136 .Y81. **44 Alamy Images:** Eddie Gerald (bc). **46-49 The Bodleian Library, University of Oxford:** MS. Pococke 375, fol. 3b-4a. **49 Getty Images:** G. Dagli Orti / De Agostini (br). **50-51 Getty Images:** De Agostini. **50 Getty Images:** British Library / Robana (cra). **52-55 Bibliothèque nationale de France, Paris. 55 Photo Scala, Florence:** White Images (br). **56-59 The Mappa Mundi Trust and Dean and Chapter of Hereford Cathedral. 59 Getty Images:** British Library / Robana (cra). **60-61 Museo Naval de Madrid. 62-65 Bibliothèque nationale de France, Paris. 65 Photo Scala, Florence:** White Images (br). **66-67 Lebrecht Music and Arts:** Pictures from History / CPAMedia. **68-71 James Ford Bell Library, University of Minnesota, Minneapolis, Minnesota. 71 SuperStock:** Iberfoto (br). **72-73 Photo Scala, Florence. 73 Corbis:** Chris Hellier (cr). **74 Getty Images:** A. Dagli Orti / De Agostini (br); De Agostini (tr). **Photo Scala, Florence:** (tl, clb). **75 Getty Images:** De Agostini (crb); G. Dagli Orti / De Agostini (cra). **Photo Scala, Florence:** (tl, bl). **76-79 Museo Naval de Madrid. 76 Getty Images:** De Agostini (cb). **79 Photo Scala, Florence:** The Metropolitan Museum of Art / Art Resource (cr). **80-81 Getty Images:** A. Dagli Orti / De Agostini. **82-83 akg-images:** Erich Lessing. **83 Corbis:** Alfredo Dagli Orti / The Art Archive (br). **84-85 Corbis. 85 Getty Images:** A. Dagli Orti / De Agostini (cra). **86-89 The Library of Congress, Washington DC:** G3200 1507 .W3 Vault. **86 University of the Americas Puebla, Mexico Foundation:** Anonymous, Martin Waldseemüller, oil on canvas, 1911 172 x 109 cm (c). **89 The Library of Congress, Washington DC:** G3201.B71 1507 .H6 1879 (cr). **90 Photo Scala, Florence:** Heritage Images (br). **91-93 Alamy Images:** Images & Stories. **94-95 Photo Scala, Florence:** British Library board / Robana. **94 Photo Scala, Florence:** BPK, Bildagentur fuer Kunst, Kultur und Geschichte, Berlin (cra). **96-99 Photo Scala, Florence:** British Library board / Robana. **99 akg-images:** Cameraphoto (cra). **100-103 © 2014, Biblioteca Apostolica Vaticana:** A.-Z.Carte Nautiche Borgiano III. by permission of Biblioteca Apostolica Vaticana, with all rights reserved. **103 akg-images:** (cra). **104-105 The Bodleian Library, University of Oxford:** MS. Arch. Selden. A1, fol. 2r. **105 Alamy Images:** Visual Arts Library (London) (bl). **106-109 Courtesy of Centre for Newfoundland Studies, Memorial University Libraries. 109 TopFoto.co.uk:** (cr). **110-113 Bibliothèque nationale de France, Paris. 110 Corbis:** Christie's Images (bc). **113 collectie Maritiem Museum Rotterdam:** (tr). **114-115 The Bodleian Library, University of Oxford:** MS. Selden Supra 105. **116-117 Getty Images:** British Library / Robana. **118-121**

Photo Scala, Florence. **121 Corbis:** Stefano Bianchetti (br). **122-125 State Library of New South Wales:** Safe / M2 470 / 1617 / 1. **122 Alamy Images:** AF Fotografie (bc). **125 akg-images:** British Library (br). **126-129 James Ford Bell Library, University of Minnesota, Minneapolis, Minnesota:** The Ricci map is owned by the James Ford. **129 Getty Images:** Hulton Archive (br). **130-133 The Bodleian Library, University of Oxford:** MS. Selden Supra 105. **130 Getty Images:** Universal Images Group (cr). **133 Corbis:** Keren Su (br). **134-137 The Library of Congress, Washington DC:** G2306.R5 M3 1644. **134 Corbis:** Imaginechina (bc). **137 The Bridgeman Art Library:** Private Collection (br). **138-141 By permission of The British Library. 141 The Art Archive:** Bodleian Libraries, The University of Oxford (br). **142-145 Harry Ransom Center, The University of Texas at Austin. 143 Het Scheepvaartmuseum:** (crb). **145 Mary Evans Picture Library:** (bc). **146-149 Gillmark Map Gallery (gillmark.com). 147 Alamy Images:** AF Fotografie (bl). **149 By permission of The British Library:** (br). **150-153 Getty Images:** British Library / Robana. **153 Corbis:** (br). **154-155 Bibliothèque nationale de France, Paris. 155 The Bridgeman Art Library:** Private Collection / Archives Charmet (tc). **Photo Scala, Florence:** White Images (tr). **156-159 The Rosenbach of the Free Library of Philadelphia Foundation:** [Haggadah]. Seder Hagadah shel Pesah: ke-minhag Ashkenaz u-Sefarad. Amsterdam: Bevet ha-meshutafim...Shalit, [5] 455 [i.e. 1695]. **159 Corbis:** Lebrecht Music & Arts (tr). **160-161 Barry Lawrence Ruderman Antique Maps / www.raremaps.com. 162-165 Bibliothèque nationale de France, Paris. 163 The Walters Art Museum, Baltimore:** Detail from Portrait by Jean-Marc Nattier 38.101 (br). **165 Alamy Images:** Image Asset Management Ltd (br). **166-167 British Geological Survey:** CP13 / 107 © NERC. All rights reserved. **168-171 By permission of The British Library. 171 Alamy Images:** Universal Art Archive (br). **172-175 The Library of Congress, Washington DC:** G3300 1775 .M5. **175 Getty Images:** British Library / Robana (br). **176 Getty Images:** Universal Images Group (cr). **176-179 Photo Scala, Florence:** BPK, Bildagentur fuer Kunst, Kultur und Geschichte, Berlin. **180-183 akg-images:** British Library. **183 Alamy Images:** AF Fotografie (br). **184-187 British Geological Survey:** CP13 / 107 © NERC. All rights reserved. **185 Science Photo Library:** Paul D Stewart (br). **187 Getty Images:** Science & Society Picture Library (br). **188-189 The Library of Congress, Washington DC:** G7960 s43 .I6. **188 Alamy Images:** amana images inc. (c). **190-191 Smithsonian Institution, Washington, DC, USA. 192-193 Photo Scala, Florence:** British Library board / Robana. **193 Wellcome Images:** (br). **194-197 The Library of Congress, Washington DC:** G3861.E9 1860 .H4 CW 13.2. **197 Princeton University Library:** Donations aux pauvres, Plate V. André-Michel Guerry. Essai sur la statistique morale de la France (1833) Historic Maps Collection, Department of Rare Books and Special Collections (br). **198-201 Royal Geographical Society. 201 Alamy Images:** AF Fotografie (br). **202-203 Smithsonian Institution, Washington, DC, USA. 203 Alamy Images:** AF Fotografie (cla). **204-207 Library of the London School of Economics & Political Science:** Booth's Maps Descriptive of London Poverty 1898-1899. **204 Getty Images:** Time & Life Pictures (cb). **207 Library of the London School of Economics & Political Science:** BOOTH / B / 354 pp6- 7 (br). **208-209 Photo Scala, Florence:** BPK, Bildagentur fuer Kunst, Kultur und Geschichte, Berlin. **208 Agence France Presse:** National Geographic (bl). **210-211 Copyright by Marie Tharp 1977/2003. Reproduced by permission of Marie Tharp. 212-215 Map Reproduction Courtesy of the Norman B. Leventhal Map Center at the Boston Public Library. 215 Royal Geographical Society** (br). **216-219 TfL from the London Transport Museum collection:** Harry Beck. **216 © 1965 Ken Garland:** (bc). **219 The Stapleton Collection:** Harry Beck (bc). **220-223 © The Estate of R. Buckminster Fuller:** Dymaxion™ Air-Ocean World, R. Buckminster Fuller & Shoji Sadao, Cartographers. **220 Getty Images:** Bachrach (bc). **223 Roland Smithies / luped.com:** (cr). **224-225 U.S. Air Force. 225 Lunar and Planetary Institute:** Image processing by Brian Fessler and Paul Spudis (cra). **226-229 Oxford Cartographers, Oxford, UK:** Akademische Verlagsanstalt, Germany / ODT, Inc. Massachusetts, USA. **229 Alamy Images:** Michael Schmeling (br). **230-231 Copyright by Marie Tharp 1977/2003. Reproduced by permission of Marie Tharp. 230 Reproduced by kind permission of Marie Tharp Maps, LLC** (c). **231 Science Photo Library:** Dr Ken Macdonald (br). **232-235 Photo Scala, Florence:** Private Collection / Photo © Christie's Images / DACS (Design And Artists Copyright Society), © Estate of Alighiero e Boetti / DACS 2014. **233 Corbis:** Christopher Felver (crb). **235 Corbis:** Finbarr O'Reilly / Reuters / DACS (Design And Artists Copyright Society), © Estate of Alighiero e Boetti / DACS 2014 (bl). **236-239 Worldmapper, © Copyright 2006 SASI Group (University of Sheffield) and Mark Newman (University of Michigan), www.worldmapper.org. 237 Getty Images:** Jeremy Sutton-Hibbert (tr). **240-243 Courtesy of TAG Fine Arts. 243 akg-images:** (cra). **244-245 Google Earth:** Map data: Google, DigitalGlobe, SIO, NOAA, U.S. Navy, NGA, GEBCO. **246-247 The Bodleian Library, University of Oxford:** MS Arab. c. 90, fols. 23b-24a

Jacket images: *Front, Back and Spine:* **Getty Images:** A. Dagli Orti / De Agostini

All other images © Dorling Kindersley

For further information see:
www.dkimages.com

About the Author

Jerry Brotton is Professor of Renaissance Studies at Queen Mary University of London, and a leading expert in the history of maps and Renaissance cartography. He is the author of *A History of the World in Twelve Maps* and *The Sale of the Late King's Goods: Charles I and his Art Collection*, which was shortlisted for the Samuel Johnson Prize and the Hessell Tiltman History Prize. He also presented the BBC television series *Maps: Power, Plunder and Possession*.